Springer Aerospace Technology

The *Springer Aerospace Technology* series is devoted to the technology of aircraft and spacecraft including design, construction, control and the science. The books present the fundamentals and applications in all fields related to aerospace engineering. The topics include aircraft, missiles, space vehicles, aircraft engines, propulsion units and related subjects.

More information about this series at http://www.springer.com/series/8613

Oleg Nicolaevich Skrypnik

Radio Navigation Systems for Airports and Airways

Oleg Nicolaevich Skrypnik
Aircraft Radio Electronic Equipment
Moscow State Technical University
of Civil Aviation
Irkutsk, Russia

ISSN 1869-1730 ISSN 1869-1749 (electronic)
Springer Aerospace Technology
ISBN 978-981-13-7200-1 ISBN 978-981-13-7201-8 (eBook)
https://doi.org/10.1007/978-981-13-7201-8

Library of Congress Control Number: 2019934521

This Springer imprint is published by the registered company Springer Nature Singapore Pte Ltd.
The registered company address is: 152 Beach Road, #21-01/04 Gateway East, Singapore 189721, Singapore

Preface

Efficiency of the air transport utilization depends significantly on the efficiency of air traffic control (ATC) system performance and is characterized by parameters such as safety, regularity, and operation economy of air traffic.

The basis of ATC system is airfield network that includes airfields of different kinds and purposes, adherent airspace areas as well as air paths. On airfields and adherent airspace areas, most complicated and challenging ATC tasks are solved such as aircraft's takeoff and landing, on the air paths—flight navigation tasks. Thereat, on all flight stages, an aircraft must be provided with the required flight safety level.

Aircraft's flight navigation and landing task solving intends highly accurate spatial position determination as well as the aircraft's motion parameters, and also precise time in order to determine the deviations from the set flight trajectory. Usage of radio-technical systems is one of the main methods of performing accurate and reliable aircraft's navigation, especially at difficult meteorological conditions and at night. That is why the whole aviation development history is followed by the development of new and continuous improvement of traditional onboard and ground radio navigational aids.

Satellite navigation systems are the most complete in functional capabilities and aircraft's position determination accuracy. According to the ICAO forecast, they will become main and sustainable mean of navigation provisioning during all flight stages, including approach and landing. However, as of now and in the midterm perspective traditional radio navigational means that are designed for aircraft's position control and its navigation in the airfield's area and on the air paths will keep its importance.

That is why it appears relevant and reasonable to review the whole traditional as well as prospective technical radio navigation aids, mounted in the airfields' areas and designed for flight navigation and landing task solving, common principles of its operation, and modern technical and constructive decisions that provide high operational rates.

In the educational book, the author tried to provide the readers with the whole picture of theoretical basis of construction, operational principles, peculiarities of usage, main characteristics, and construction of radio navigational systems that are mounted in airports. As basic characteristics for consideration, constructions and circuitry systems made in Russia were taken. Thereat, the author tried to illustrate some peculiarities in terminology and system approach that add some particularity to Russian civil aviation.

This educational book is written for students of aviation department and universities. It can also be useful for pilots and specialists who operate ground and onboard radio navigational flights support means.

Irkutsk, Russia Oleg Nicolaevich Skrypnik

Acknowledgements

The author thanks Karina Skrypnik (Preface, Abstracts, Summaries, Chaps. 1, Sects. 3.1–3.2) and Tatyana Portnova (Chaps. 2, 4, 5, 6, Sect. 3.3) for providing their support in translation and also Roman Aref'ev who took part in research conduction, which results are described in Chap. 5, and Gleb Mal'ko who took part in research conduction, which results are described in Chap. 6.

Contents

Abbreviations

ACP	Alternate Command Post
ACR	Airfield Control Radar
ADC	Analog–Digital Convertor
ADF	Automatic Direction Finder
ADN	Antenna Distribution Network
AM	Amplitude Modulation/Modulator
AMU	Audio Management Unit
AP	Antenna Pattern
APCH	Approach
ASR	Airport Surveillance Radar
ATA	Actual Track Angle
ATC	Air Traffic Controller
ATCT	Air Traffic Control Tower
ATN	Aeronautical Telecommunication Network
AWGN	Additive White Gaussian Noise
BFO	Beat Frequency Oscillator
BM	Balanced Modulator
BPA	Box of Processing and Automatics
BPSK	Binary Phase-Shift Keying
B-RNAV	Basic RNAV
BVOR	Broadcast VOR
C/A	Clear Access (coarse/acquisition)
CDI	Course Deviation Indicator
CDMA	Code-Division Multiple Access
CEA	Curve of the Equal Angles
CF	Carrier Frequency
CNS/ATM	Communication, Navigation, Surveillance/Air Traffic Management
COP	Circle of Position
CSB	Carrier and Side Band
CSP	Command-the-Starting Point

CTP	Control Tower Point
CVOR	Conventional VOR
DA	Drift Angle
DAC	Digital-to-Analog Converter
DACC	Device for Antenna Circuit Control
DCSF	Device for Call Sign Formation
DDM	Difference in Depth of Modulation
DF	Direction Finder
DH	Decision Height
DME	Distance-Measuring Equipment
DoSM	Depths of Space Modulation
DR	Dead Reckoning
DSP	Digital Signal Processor
DVOR	Doppler VOR
E	Engine
emf	Electromotive Force
EG	Electric Goniometer
EGM	Earth Gravitational Model
EGNOS	European Geostationary Navigation Overlay Service
eLoran	Enhanced Loran
ESD	Equisignal Directions
EV	Expected Value
FA	Final Approach
FAA	Federal Aviation Administration
FDMA	Frequency-Division Multiple Access
FIS	Former of Information Signals
FL	Flight Level
FM	Frequency Modulation/Modulator
FMGC	Flight Management and Guidance Computer
FMS	Flight Management System
FTE	Flight Technical Error
G/S	Glide Slope
GAGAN	GPS-Aided GEO Augmented Navigation
GBAS	Ground-Based Augmentation Systems
GCA	Ground-Controlled Approach
GDOP	Geometric Dilution of Precision
GEO	Geostationary Earth Orbit
GLONASS	GLObalnaya NAvigacionnaya Sputnikovaya Sistema
GLS	GBAS Landing System
GNSS	Global Navigation Satellite System
GPS	Global Positioning System
GRAS	Ground-based Regional Augmentation System
GRS	Geodesic Reference System
GSO	Geosynchronous Orbit
HDOP	Horizontal Dilution of Precision

HFG	High-Frequency Generator
HP	High Precision
HSI	Horizontal Situation Indicator
IA	Initial Approach
ICAO	International Civil Aviation Organization
ICNIA	Integrated Communication, Navigation, Identification Avionics
IERS	Earth Rotation and Reference Systems Service
ILS	Instrumental Landing System
IM	Inner Marker
IRS	Inertial Reference System
ITRF	International Terrestrial Reference Frame
ITRS	International Terrestrial Reference System
LAAS	Local Area Augmentation System
LCD	Liquid Crystal Display
LEDD	Line of the Equal Differences in the Distance
LF	Low Frequency
LFA	Low-Frequency Amplifier
LFG	Low-Frequency Generator
LKKS	Lokal'naja Kontrol'no-Korrektirujushhaja Stancija
LMM	Locator Middle Marker
LMSL	Local Mean Sea Level
LOC	Localizer
LOM	Locator Outer Marker
LOP	Line of Position
LRA	Low-Range Altimeter
MB	Marker Beacon
MEO	Medium Earth Orbit
MF	Medium Frequency
MFBARS	Multifunctional Multiband Airborne Radio System
MLAT	Multilateral System
MLS	Microwave Landing System
MM	Middle Marker
MMR	Multimode Receiver
MRR	Marker Radio Receiver
MSAS	Multifunctional Satellite-Based Augmentation System
MSE	Mean Squared Error
MSL	Mean Sea Level
NAD	North American Datum
NAVD	North American Vertical Datum
NAVSTAR	NAVigation Satellites providing Time And Range
NBF	Narrow-Band Filter
ND	Navigation Display
NDB	Non-Directional Beacon
NSE	Navigation System Error
OM	Outer Marker

OSP	Oborudovanie Sistemy Posadki
P	Precise (protected)
PA	Power Amplifier
PAR	Precision Approach Radar
PBN	Performance-Based Navigation
PD	Phase Discriminator
PDE	Position Definition Error
PDOP	Position Dilution of Precision
PL	Pseudo-lite
PRN	Pseudo-random Noise
P-RNAV	Precision RNAV
PZ	Parametry Zemli
QFE	Question Field Elevation
QNE	Question Nautical Elevation
QNH	Question Nautical Height
RB	Radio Beacon
RBI	Relative Bearing Indicator
RCE	Remote Control Equipment
RCVR	Receiver
RLS	Radar Landing System
RMI	Radio Magnetic Indicator
RMS	Radial Mean Square Error
RNAV	Area Navigation
RNP	Required Navigation Performance
RNS	Radio Navigation System
RS	Reference Station
RSBN	Radiotehnicheskaja Sistema Blizhnej Navigacii
RSP	Radiolokacionnaya Sistema Posadki
RVCG	Reference Voltage-Controlled Generator
RVG	Reference Voltage Generator
RVR	Runway Visual Range
RW	Runway
SBAS	Satellite-based Augmentation System
SBM	Single-Band Modulator
SBO	Side Band Only
SCFO	Subcarrier Frequency Oscillator
SDB	Sum-and-Difference Bridge
SDKM	Sistema Differencial'noj Korrekcii i Monitoringa
SF	Side Frequency
SHF	Super-High Frequency
SK	Sistema Koordinat
SNS	Satellite Navigation System
SOF	Surface of Position
SP	Squitter Pulses
SQM	Signal Quality Manager

SRRNS	Short-Range Radio-technical Navigation System
SST	Signal Shaper and Tester
SV	Space Vehicle
TA	Transition Altitude
TACAN	TACtical Air Navigation system
TDOP	Timing Dilution of Precision
TG	Tone Generator
TH	Transition Height
TL	Transition Layer
TO	Test Oscillator
TRB	True Radio Bearing
TSE	Total System Error
TVOR	Terminal VOR
TX	Transmitter
UCSD	Unit of Computing and Switching Devices
UHF	Ultra-High Frequency
UTC	Coordinated Universal Time
VDF	VHF Direction Finder
VDL	VHF Datalink
VDOP	Vertical Dilution of Precision
VFR	Visual Flight Rules
VHF	Very High Frequency
VOR	VHF Omnidirectional Radio Range system
WA	Wind Angle
WAAS	Wide Area Augmentation System
WGS	World Geodetic System
WP	Way Point

Chapter 1
Elements of the General Radio Navigation Theory

This book is dedicated to the overview of ground-based air radio navigation aids. These aids are part of radio navigation systems and operate together with onboard equipment. Principles of construction, functioning, structural, and circuit design for ground and onboard equipment are very different. However, it is the interoperability of onboard and ground equipment of radio navigation systems that allows to solve a wide range of tasks required for completing all stages of aircraft's flight with the necessary level of flight security.

The fundamental operating principle of radio technical navigation means is patterns and peculiarities of radio waves propagation. With the help of specially developed methods and technical aids these peculiarities are used to form the navigation information on the transmitting side as well as its extraction on the receiving side. These methods as well as the main terms and definitions, ways of mathematical description of aircrafts' navigation principles, processes of formation, transmission, and processing of navigation information make the bottom line of common radio navigation theory.

Radio navigation theory is relatively recent but already formed science. Thereat, the radio navigation theory is considered a developing science which is defined by new achievements of humanity in related areas, first of all—in the area of satellite technologies and digital signals processing.

Knowledge of radio navigation theory basis which is described in this chapter will allow to better understand the principles of construction and functioning of radio navigation systems which are covered in the following chapters of this book.

Learning material of this chapter is presented in the following sequence:

Section 1.1 gives the overview on common principles of air navigation theory, navigation tasks, aircraft's flight stages as well as on technical aids used for navigation, its advantages, and disadvantages. Basic tactic and technical characteristics of radio navigation aids are described, and its meaning is explained.

Section 1.2 describes main characteristics and principles of radio-wave propagation, which allows its use in radio navigation systems as well as the main principles of radio navigation systems construction. This section also gives the classification of radio navigation systems on various properties. Structural diagrams are overviewed,

O. N. Skrypnik, *Radio Navigation Systems for Airports and Airways*,
Springer Aerospace Technology, https://doi.org/10.1007/978-981-13-7201-8_1

and advantages and disadvantages of systems with and without transponder and multiposition systems are shown.

In Sect. 1.3, main navigation terminology is described, their meaning is explained, and their graphical interpretation is given. Some special aspects of aviation terminology used in Russia are mentioned.

Section 1.4 gives an overview on coordinate systems used in air navigation. Special aspects of local, global and aircraft-related (with aircraft principal axes) coordinate systems are shown, their graphic interpretation is given. Coordinate transformations for conversion the object's coordinates from one coordinate system to another are given. Evolution stages global coordinate systems from reference ellipsoids to its modern look are described. Coordinate systems recommended by ICAO as international as well as the coordinate systems used in Russia are shown. Differences between these coordinate systems are described.

Section 1.5 describes the surfaces and lines of position which are formed by radio navigation systems. Its graphic interpretation is given as well as the number of formulas used for navigation calculations.

Section 1.6 describes the methods of position determination used in the air navigation; their advantages and disadvantages are shown. Method of getting the equation for position determination error by the method of lines of position—for radial mean square error (RMSE)—is given. Conditions at which the minimal RMSE is achieved are given.

Section 1.7 describes coverage areas (working zones) of main types of radio navigation systems. For each of described systems, method of getting the formulas used for constructing the coverage area is given. Section contains images that show coverage areas of direction-measuring (theta) system, range (rho)-measuring system, direction-range (rho-theta) system, and range difference (hyperbolic fixing) system; configuration and coverage areas' size changes at different conditions are overviewed.

Section 1.8 gives an overview on the main requirements of aviation consumers to the navigation accuracy in accordance with the performance-based navigation (PBN) concept. The meaning of navigation specifications RNAV and RNP are explained. ICAO recommended specifications for different stages of flight are given.

1.1 Air Navigation and Its Technical Aids Overview

The term "navigation" comes from the Latin word "navigation"—swimming on board. During a long period of human evolution (starting from the fifth century before our era when first maps in a form of plans and first technical navigation aids—magnet compasses (China) appeared, till the twentieth century, when new objects of navigation and new meanings of this term emerged), the evolution of navigation methods and tools was defined by necessity of solving problems in marine navigation.

Starting from the mid of the twentieth century, the evolution of technical aids of navigation began to be defined by aviation and a new course in navigation appeared—the air navigation.

In the modern interpretation, the term "navigation" means the process of controlling an object moving in a certain space, and it is also science about methods and tools of determining the position and parameters of an object moving in this space. Thereat, the space of an object moving might be the physical environment (earth and sea surface, undersea and under-earth, aerial and space environment), as well as the virtual (e.g., informational space) and the navigation objects might be technical, biological, and other systems. According to this, such types of navigation can be identified—automotive, nautical, aerial, space, inertial, informative, bionavigation, and other types—which are different by their methods and the technical tools of problem solving.

Air navigation is the science about methods and tools of receiving the information about the current aircraft's motion position and parameters and about methods and tools of navigation at current position and trajectory of airspace's motion uncertainty. Thereat, by the flight path we understand the curve in the airspace along which the aircraft (its center of mass) is moving.

The main role of air navigation is leading the aircraft by the optimal trajectory to the target point or area at the given time. Thereat, by the optimal trajectory we understand such trajectory which provides the required level of flight security in existing air situation conditions. It is the air situation that leads to optimal trajectory of a flight not always being the shortest distance between its initial and end point.

The resolution of the main navigation's task is divided into the number of particular tasks different by their mode and methods of resolution. The particular tasks include:

selection and calculation of a flight's path optimal trajectory and the aircraft's motion parameters (during planning and in flight due to the air conditions change);
measurement of the aircraft's motion main navigation parameters, values that describe current position, direction, and its motion speed;
comparison of the results of defining navigation parameters with given or calculated values and corrective actions production that provide the aircraft motion on the optimal trajectory.

The process of aircraft motion support is divided into two interrelated tasks: the navigation itself (the navigation task of moving its center of mass on the predefined spatiotemporal trajectory) and piloting—control of aircraft's angle position in airspace. Due to the peculiarities of navigation provisioning, the navigation task can be divided into such main phases like: en route flight (including oceanic/remote areas), flight in terminal that includes takeoff, and approach to landing (landing in non-precision approach and landing in precision approach) meeting the ICAO requirements.

The whole aircraft's navigation process can be divided into separate sequential milestones (Fig. 1.1):

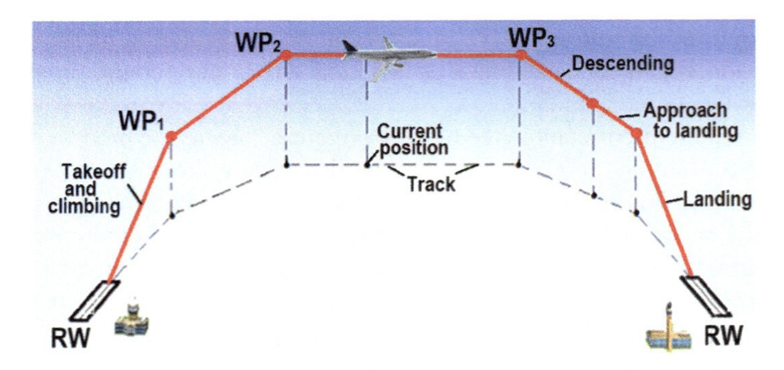

Fig. 1.1 Sequential milestones of aircraft's navigation process

takeoff and climbing;
entering the way point (WP);
the flight by the route specified via transitional way points with controlling and correction of deviation from the route;
entering the target—runway (RW) for example;
descending approach (guidance to the threshold of RW) performing all the necessary maneuvers and landing.

Depending on the air situation, peculiarities of an aircraft usage, and air traffic management in the usable airspace, additional steps of navigation process might emerge. For instance, routes of so-called area navigation (RNAV) that allow to conduct a flight on the straight, omitting way points were imposed. In the medium term, the migration to free (flexible) flights enabling the flight crew of an aircraft to choose the optimal flight trajectory is possible. The increasing value due to the flight safety requirements obtains the surface operations which include navigation on the airport surface (taxiways) to and from the active runway.

By characterizing the aircraft navigation, 2D, 3D, and 4D navigation types are distinguished.

2D navigation considers setting the route and coursing only the flight path, for what the information of the horizontal coordinates of an aircraft is required. In 3D navigation, the flight altitude/height setting and its control are added. In 4D navigation, besides setting and control of three-dimensional coordinates, the strict binding of the aircraft flight trajectory (time of passing the way points) to time is required.

To identify the navigation parameters, different air navigation technologies are being used, which can be divided into five main groups by physical principles of mode operation used in them:

1. Navigation and piloting means, based on the usage of the geophysical characteristics of the Earth and its atmosphere (geomagnetism, atmosphere pressure changing with the altitude, correlation of the air pressure and speed, etc.). Magnetic compasses, air speed indicators, and barometric altimeters fit to this group.

They are easy to use and reliable; however, they have low accuracy and solve limited number of navigation tasks.
2. Lighting and other visual aids to air traffic support (ground lights, runway lights, spotlights, color markers, signal rockets, etc.). These aids can be used only at good visibility and help you solve a very limited number of navigation tasks, connected with the flight crew spatial orientation.
3. Astronomic navigation aids, based on the usage of the regular change of the relative position of Earth and different celestial bodies. Astronomic compass, aviation sextant, automatic astro-tracker fit to this group. These aids are difficult to use and have low accuracy, and its usage is possible only at good visibility of celestial bodies.
4. Inertial navigation aids, based on the aircraft's acceleration measurement caused by the non-gravitational forces (engine thrust, ascensional power). Integration of accelerations, measured on board of an aircraft, let you define components of its speed and the following integration of components—position coordinates. Inertial navigation systems, gyrovertical and air data system fit to this group.

Advantages of inertial tools are endurance and high reliability, continuous information output about aircraft's coordinates and motion parameters. However, inaccuracies in aircraft's speed and coordinates reckoning by using the inertial navigation tools accumulate with time which is the big disadvantage of these tools. Besides, inertial navigation systems are high-technology products which determine its high cost.

5. Radio-technical navigation aids, based on the usage of consistency and radio waves conditions expansion in the terrestrial and cosmic space.

Radio-technical aids provide high navigational parameters' measurements accuracy at any weather conditions and let you solve various and complicated navigational issues, they can have long operational range. However, these aids are exposed to radio interference and are not autonomous in most cases. Besides, the accuracy of position determination by radio navigation systems often depends on the mutual aircrafts' and the radio beacons position which create navigation-temporary field and the wave propagation conditions.

Radio-technical navigation tools' implementation has various concrete forms such as navigation devices, navigation systems, complex navigation systems, and multifunctional integrated systems.

Radio navigational devices are the airborne autonomous measuring elements (radio altimeters, Doppler speed and drift angle-measuring devices, radio altimeter systems).

Radio navigation systems consist of airborne and ground-based navigation aids (range- and direction-measuring radio navigation systems, radio-technical landing systems, short- and long-range radio navigation systems, satellite navigation systems). Airborne hardware usually provides the signal detection and its processing, and the ground-based (radio beacons)—the forming of the radio signals of required structure and its emission. In the satellite navigation systems, the aids that are forming and emitting the radio signals are located on the artificial Earth satellites.

Airborne complex navigation systems usually include navigation aids and systems which measure the same navigation parameters, but which work based on different physical principles. As an example, it is the radio navigation measurers and inertial navigation system incorporation, where its output information is processed in a flight management computer system (FMCS) with the help of special algorithms of complex information processing. This complexity, based on the redundancy of information about measured parameter, helps to increase the accuracy and reliability of navigational sighting.

Lately, the elaboration focused on the creation of ground-based complex radio navigation systems of airfields started to appear. In such systems, the complexing effect is in the maximum unification of navigation, surveillance, and landing systems by the functional modules apparatus, constructive performance, and supreme technical solution standardization of internal and external interfaces, united software and unified maintenance, control and management. As an example of such system, there is a unified ground-based landing systems' radio installation, navigation, and air traffic management for the magistral and local airlines "Complex 734" airfields (LLC Research and development company «Radio engineering systems», *NPO "Radiotekhnicheskie sistemy"*, Russia).

Airborne multifunctional integrated systems are the next step of complex navigation systems development. The built-in and functional integration is used in these systems, which helps not only to increase the accuracy and reliability of navigation parameters determination, lower the weight, dimensions, airborne hardware power consumption, but also to get the qualitative new functions. For better efficiency of tasks solving during various flight stages, the integrated system can have flexible architecture and controlling resources of flight navigation computing system. One of examples is the well recommended in practice inertial-satellite navigation system, uniting two subsystems into one module. Other examples are the military aviation accepted concepts—Integrated Communication, Navigation, Identification Avionics (ICNIA), Multifunctional Multiband Airborne Radio System (MFBARS), and Communication, Navigation, Surveillance/Air Traffic Management (CNS/ATM) for the civil aviation.

Nomenclature of ground-based radio navigation aids, used for solving navigation and landing tasks in the airfield's area, is pretty diverse. This is connected not only with the peculiarities of air navigation provision in the airfield's area but also with the peculiarities of navigation parameters' measurements with the help of radiotechnical methods. Besides, lately the necessity of aircraft's position control when in motion on the airfield has arisen. For solving this task and also for takeoff and the final stage of aircraft descending control, so-called multilateration systems (MLAT) were created. Their peculiarity is in the usage of not only the radio navigation, but also the radio location methods and appropriate technical aids for solving the classic navigation task to determine aircraft's coordinates.

The ground-based radio navigation airfields' equipment includes:

automatic VHF direction-finding equipment (VDF), that enables you to determine direction on the aircraft by the onboard VHF radio station's emission;

non-directional beacons (NDB), which signal acquisition by onboard automatic direction finder enables you to determine the direction to NDB;
marker radio beacons (Marker), which signal acquisition by onboard radio station enables you to determine the fact of an aircraft passage over the marker beacon;
range scope transponders, receiving and then retranslating the signals from the onboard interrogators, acquisition of which by onboard equipment enables you to determine the distance to the range scope transponder ***(distance-measuring equipment, DME)***;
radio beacons of short-range VHF band navigation radio systems, emitting the signals, which acquisition by onboard equipment enables you to determine the direction to the radio beacon on the board ***(VHF omnidirectional radio range system, VOR)***;
radio beacons of distance- and direction-measuring (rho-theta) radio-technical short-range systems of UHF band, signal acquisition of which by onboard equipment enables you to determine the aircraft's direction and distance toward the radio beacon (***TACAN*** (USA) and ***RSBN*** (Russia) system).
radio-technical landing systems' radio beacons, that are emitting the signals, acquisition of which by onboard equipment enables you to determine the value and direction of aircraft's deviation toward the defined approach trajectory as well as the distance to the runway ***(instrumental landing system, ILS)***.

The possibilities of practical radio-technical navigation aids realization and their function are determined by the combination of tactical and technical characteristics. Tactical system characteristics are the ones that determine its functional capabilities when applied. Technical system characteristics reflect engineering solutions that provide accomplishment of the defined tactical characteristics.

Main ***tactical characteristics*** include: navigation parameters determination accuracy, coverage area, signal propagation, availability, integrity, maintenance continuity, position determination intermittency (fix rate), system capacity and ambiguity.

The coverage (working) area "provided by a navigation system is that surface area or space volume in which the signals are adequate to permit the user to determine position with a specified level of accuracy. Coverage area is influenced by system geometry, signal power levels, receiver sensitivity, atmospheric noise conditions, and other factors that affect signal availability" [1].

With the growth of air traffic's intensity and density as well as the speed, altitudes and route-length increase; higher requirements to navigation equipment are demanded. This predefined the necessity of requirements' satisfaction for creating the conditions of accurate position determination from any Earth point and terrestrial space, i.e., the global coverage area requirements.

"In navigation, ***the accuracy*** is the degree of conformance between the estimated or measured parameter (position, velocity and other) of a platform at a given time and its true parameter.

Since accuracy is a statistical measure of performance, a statement of system accuracy is meaningless unless it includes a statement of the uncertainty in position that applies" [1].

Requirements for the accuracy of aircraft's navigation parameters determination depend on its function and the type of tasks it can solve, as well as on the category of airspace (oceanic area, internal continental route, or special problems accomplishment area).

"***The availability*** of a navigation system is the percentage of time that the services of the system are usable. Availability is an indication of the ability of the system to provide usable service within the specified coverage area. Signal availability is the percentage of time that navigation signals transmitted from external sources are available for use. It is a function of both the physical characteristics of the environment and the technical capabilities of the transmitter facilities" [1].

Availability is defined by the probability of getting the authentic navigation-temporal information system within its working area and at specified time and with defined accuracy.

Availability requirements change depending on the phases of flight and the tasks, solved by an aircraft. In terms of an aircraft's flights safety provision, the highest requirements, at which the availability must be equal to 1, are specified by approaching and landing by ICAO categories.

"***The integrity*** is the measure of the trust that can be placed in the correctness of the information supplied by a navigation system. Integrity includes the ability of the system to provide timely warnings to users when the system should not be used for navigation" [1].

Integrity is defined by its corresponding probability.

System integrity requirements for air consumers (users) are the highest because of the high speed of an aircraft and infeasibility of large gaps of information updating.

The continuity of a system is the ability of the total system (comprising all elements necessary to maintain aircraft position within the defined airspace) to perform its function without interruption during the intended operation. More specifically, continuity is the probability that the specified system performance will be maintained for the duration of a phase of operation, presuming that the system was available at the beginning of that phase of operation.

Maintenance continuity is defined by its corresponding probability.

System capacity is the number of users that a system can accommodate simultaneously or at a time unit.

Taking into account the significant importance of up-to-date navigation information acquisition for providing the flight safety, system capacity must be unlimited while the continuity, i.e., the reliability of maintenance, must correspond the defined value.

Signal propagation describes the max distance $D_{\max}$, at which the signal acquisitioning acquires the minimal acceptable level $P_s = P_{s\min}$, which is yet enough for accomplishing the main functions of a system with the qualitative criteria, not worse than predefined ones.

The signal propagation depends on the radio waves band being used, as well as on the capacity of transmitting and acuity of receiving devices, and also on the directional properties of antenna systems.

Resolution ratio (ambiguity) describes the ability of discrete detection and measure of nearby objects' navigation parameters.

"System ambiguity exists when the navigation system identifies two or more possible positions of the aircraft, with the same set of measurements, with no indication of which is the most nearly correct position. The potential for system ambiguities should be identified along with provision for users to identify and resolve them" [1].

Resolution ratio is distinguished by its range, angular coordinates, and speed.

"***The fix rate*** is defined by the number of independent position fixes or data points available from the system per unit time" [1].

Noise immunity describes the ability of reliable tasks performing at natural or man-made disruptions. Interference resistance is defined by the system emission security and its resistance. System emission security describes the complexity of its performance detection and measurement of radiated signal parameters, which leads to special jamming formation.

"***The reliability*** of a navigation system is a function of the frequency with which failures occur within the system. It is the probability that a system will perform its function within defined performance limits for a specified period of time under given operating as well as storage and transportation conditions. Formally, reliability is one minus the probability of system failure" [1].

Main ***technical features*** include the method of measuring the navigation parameters, operational frequencies and their stability, emission capacity, signal modulation type, weaves bandwidth, form and width of antenna pattern, antenna gain, receiver's bandwidth and sensitivity, as well as the display components and information retrieval's characteristics, dimensions, mass and energy, consumed from power supplies.

The combination of technical characteristics provides the set tactical and operating demands that are made to the system. Since the correlation between operating and technical parameters is ambiguous, most rational and efficient ways of solving the task of correlation determination are being found in the process of system developing.

1.2 Main Building Principles and Classification of Radio Navigation Aids

At the heart of radio-technical navigation aids operation are the logic and features of radio-wave transmission in the terrestrial space, main of which are:

linearity and high stability of radio-wave transmission speed in free space and homogeneous environment;
ability to reflect radio waves from the earth's surface and other objects;
effect of receives signals' frequency drift toward the radiated signals' frequency, which appears at the mutual radio-wave source and receiver's motion (***Doppler effect***).

All radio-technical methods of navigation parameters' determination (distance, speed, direction of arrival, etc.) use functional dependence between radio signals parameters (amplitude, phase, frequency, and time of signal propagation along the radio path) and navigation parameters' values. For instance, the phasic raid is proportioned to the distance travelled by radio signal; signal amplitude on the pickup-antenna output, which has the directional properties, depends on the direction of signal emission source; time of signal propagation along the radio path is proportioned to the length of this radio path. Therefore, radio signals that are used in the navigation systems are physical carrier of navigation information.

Radio navigation systems belong to the systems of information extraction since their main function is acquisition of variable data about coordinates, speed, motion direction, and aircraft's space orientation by processing the received signals.

From the main phases of transmission, processing, and acquisition of the navigation's information perspective, any radio navigation system is built on the "with transponder" or "without transponder" system principle or as multiposition system.

In "***without transponder***" system signals are transmitted only in one direction on one radio channel ("Earth-aircraft" or "aircraft-Earth") (Fig. 1.2).

The system consists of transmission (TX) and receiving (RCVR) devices with antennas, measuring device, the navigation information output of which comes to user (onto the information display for example), as well as the reference parameter generator, used for navigation information extraction. Reference parameters can be the phase of radiated signal, its frequency, the moment of emission. Transmission device is the apparatus of high-frequent signal (including its modulation and coding for increasing the jamming resistance of data transmitting and processing) structure forming, and the receiving device is the apparatus which provides the information signals extraction and modification (demodulation and decoding) to the mode, suitable for delivery to the measuring device of system.

The reference parameter (frequency, phase, time of signal emission) must be formed both on the receiving and on the transmitting side of the system. While using highly stable and mutually synchronized in both locations reference parameter generators (time and frequency standards) for forming the reference parameters (frequency, phase, time), the measurement of any navigation parameters is possible.

The advantages of "without transponder" system include the usage of single radio channel which allows to spare the frequency resource, radio stealth, since the object

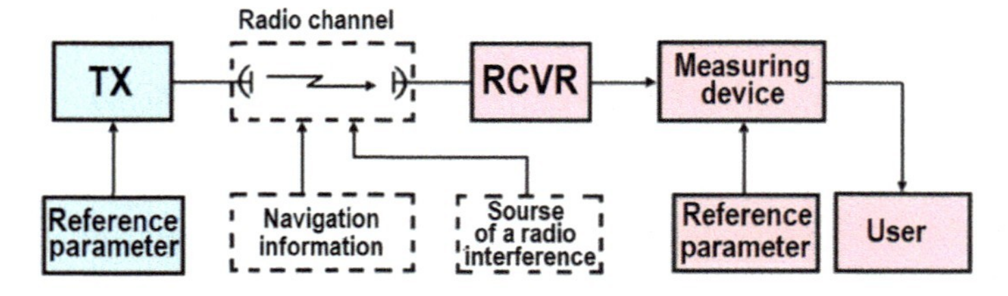

Fig. 1.2 Block diagram of the "without transponder" system

being determined is not operating on the emission, unlimited system capacity. Main disadvantage of the system includes the necessity of creating the reference parameter on the receiving side which is equivalent to the reference parameter on the transmitting side. In practice, this makes quite a complicated task to perform.

In "***with transponder***" system (with signal retranslation) (Fig. 1.3) signals are transmitted in two directions ("Earth-aircraft-Earth" or "aircraft-Earth-aircraft"). For systems, using reflected signals, an object reflecting the signal should be considered instead of transmitter–receiver. Earth surface is an example of such object.

Advantages of "with transponder" system include creation and storage of reference parameter on the determining object, which excludes the necessity of system's both parts synchronization by reference parameter. In case when the transponder is the earth station, the possibility to provide good energetic characteristics of a radio channel response arises.

Main disadvantages of the system include the necessity of dividing request's and response's channels by frequency which spends the frequency resource of system, lack of reticence, and limited system capacity since the process of receiving the requested and forming the responding signal takes some time during which other requests are not processed.

Multiposition system can use both "with transponder" and "without transponder" operating principles. Its main distinguishing feature is existence of a number of earth stations located in the points with certain coordinates (positions). Navigation task solution is accomplished either by onboard apparatus of an aircraft by receiving and coprocessing the signals, radiated either by earth stations or on the ground point via navigation parameters measurements coprocessing that are received on the earth stations. In the former case, the mode of earth station emission needs to be regulated (synchronized by its reference parameter). In the latter case, synchronization on earth station's reference parameter is necessary as well. In practice, multiposition radio navigation systems use "without transponder" operating principle more frequently (Fig. 1.4).

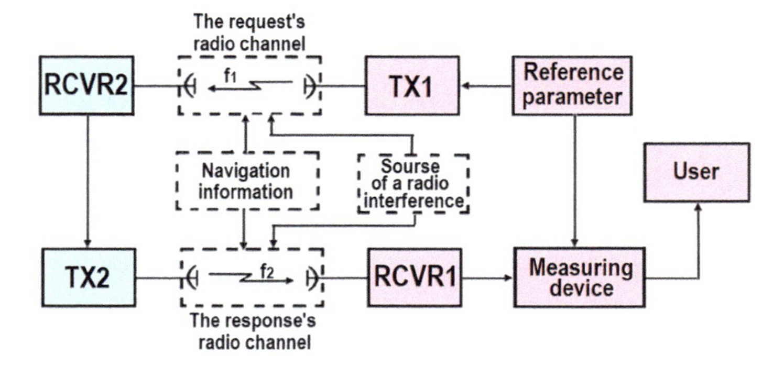

Fig. 1.3 Block diagram of the "with transponder" (with signal retranslation) system

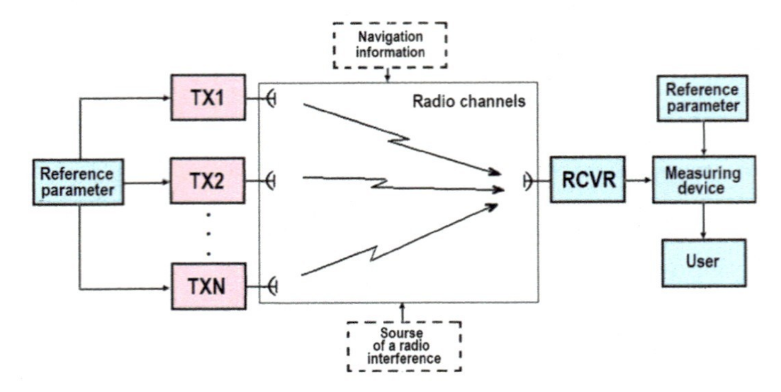

Fig. 1.4 Block diagram of the "without transponder" multiposition system

In general, radio signals radiated by the transmitter do not contain navigation information and are made for correlation of events in various points of navigation-temporal field. The navigation's information place of origin is the radio channel, during expansion along which radio signals parameters change as a result of length change, radio lines' orientation in space, mutual motion of radio signals' transmitter and receiver. Therefore, radio navigation systems in contrast with radio communication systems, which form the information as is in the transmitter by inner modulation of signals, are called systems with an external modulation of signals.

The measuring device, in which all the receiver's information signals' output parameters are compared to the reference parameter, setting the starting point of measured data, is the place of navigation information forming in the usable way.

In common case, the navigation radio signal is the function of number of parameters, each of which carries the information about aircraft motion parameters. Besides, radio navigation systems are multiposition systems by its construction principle. This way, navigation signal is a multiparameter information storage item, for acquisition of which the multidimensional processing must be carried out.

Main problem which arises in the process of radio navigation systems' development and usage is radio channel jamming resistance increase and navigation information extraction accuracy.

For navigation information extraction, radio signals' parameters must be evaluated under the conditions when there are interference in radio channel and noises in apparatus of receiving and processing, which disrupt the navigation information. Therefore, the whole process of navigation information extraction can be divided into the number of tasks that correspond with the principles of parameters evaluation. If the decision about presence or absence of useful signal in the receiver's input, the detection task occurs. If the belonging of received signal to the concrete source of emission is evaluated (from the whole combination of received signals), the recognition or signal distinction task occurs. These two tasks are usually solved in the "***search***" mode of the system. If the task of signals' detection and distinction is already solved, then the other task, the task of information parameter evaluation (or the navigation information receiving), occurs. The evaluation task is solved in the "***tracking***" mode of the system.

However, independently from all the in-between tasks and phases of information processing, the ultimate aim of radio navigation system's operation is the navigation parameter's measurement result generation.

Radio navigation systems can be classified by the following main features:

by the type of navigation parameter being measured;
by the type of radio signal parameter being measured, that is used to determine the navigation parameter;
by the methods of aircraft's position determination;
by its unction;
by range coverage and other features.

By the navigation parameter measured type radio navigation systems are divided into direction-measuring (radial, theta), distance-measuring (ranging, rho), differential distance-measuring (hyperbolic fixing), compound (enabling to determine a number of navigation parameters, for example, the direction and distance to the source of emission, distance to the source of emission and the speed of its change) systems, as well as into linear and angle rate-measuring instruments.

By the type of radio signal measured parameter amplitude, phase, frequent, and temporary (impulse) radio navigation systems are distinguished. In order to increase systems' efficiency, various radio signal parameters (amplitude-phase, impulse-phase, and other methods of measurements) can be measured simultaneously.

By position determination methods radio navigation systems can be divided into positional (based on the line or surface position's determination, matching the measured values of navigation parameters), map matching, and dead reckoning.

By function navigation, landing, collision avoidance and dangerous closing, interaircraft, and relative navigation, en route air navigation systems are described.

By the range of action, short-range navigation (that enable to determine the position of an aircraft 400–500 km away from the ground-based beacon within line of sight), long-range navigation (that enable to determine the aircraft position up to 3,000–3,500 km away from the ground-based radio navigation station), and global navigation (unlimited range of action) systems are distinguished.

Beside this, radio navigation systems like any radio electronic measuring systems can be divided according to some other features:

by nature of emission—with continuous (modulated and non-modulated), impulse (with large cycle), and continuous-impulse emission (with small cycle), as well as with the emission of signals, which parameters (frequency, phase) change by pseudo-random sequence rule (so-called pseudo-random noise (PRN) signals);
by the order of endurance—autonomous and non-autonomous;
by the order of automatization—automatic, semi-automatic, and non-automatic;
by the indication method—with visual (pointer instrument, digital display, electron-ray tube, LCD) and with aural indication.

1.3 Main Navigation Notions and Terms

Navigation parameters describe the center-of-mass position of an aircraft in space and toward the radio navigation points (e.g., ground beacons), its motion as well as its direction from an aircraft to beacons or vice versa, from beacons to an aircraft.

Navigation parameters include:

position (coordinates);
altitude and speed parameters that describe the aircraft motion toward the atmosphere or Earth surface (air, vertical, ground speed);
directions (bearing, relative bearing, heading, track angle).

Pilot-centered (piloting) parameters describe the aircraft motion toward its center of mass.

They include:

angular orientation of aircraft's main axes toward the Earth axis system (roll, pitch, and course);
angular orientation of an aircraft toward the velocity vector (angle of attack, gliding, and drift).

The list of navigation and piloting parameters is quite long. Below those parameters that will be mentioned in the following chapters of this course book will be analyzed.

Current position is the projection of an aircraft's center of mass to the Earth surface. It is determined by the coordinates connected to the Earth surface's system of axis.

Flight height is determined as a distance from the defined reference level, a point or an object considered as a point, that is connected to the Earth surface, to the aircraft, which is vertically measured.

For flight height measurement, special devices that are called altimeters are used on board.

In Russia and some other countries, the height is measured and indicated in meters, while in most countries it is measured in pounds (1 lb = 0.305 m).

Height stands for the vertical distance of a level, a point, or an object considered as a point, measured from a specified reference level.

Height $\boldsymbol{H}_{true}$ (called ***True Height*** in Russia) (Fig. 1.5) is measured using radio altimeters and features the distance from the lowest point of fuselage to the earth surface.

Other altitudes are measured using the barometrical altimeters by the level of atmosphere pressure on the aircraft's flight trajectory. Strictly speaking, measured barometrical altitude is not the physical distance; it features the barometrical altimeter indications toward the set level of atmosphere pressure (isobaric surface), taken for the reference point.

Reference point, where the barometric altimeter starts to reckon the altitude, is defined by pilot during the flight by setting the value of atmosphere pressure on the altimeter, when it must show zero.

Altitude H_a (called ***Absolute Altitude*** in Russia) (Fig. 1.5) stands for the flight altitude, measured toward the ***mean sea level (MSL)***, that matches the Geoid surface (same level surfaces of gravitation force potential) in that location.

In aviation, the absolute altitude, measured by the pressure toward the isobaric surface, which matches standard atmosphere at the sea level, is called ***Question Nautical Height (QNH)***. At setting this pressure on the altimeter, the device will indicate the aircraft flight altitude toward the sea level. For determining the real aircraft's flight altitude over the sea level, using this type of pressure, only the excess of terrain relief above sea level is required.

Height H_r (called ***Relative Altitude*** in Russia) is the altitude, measured toward the defined level on the Earth surface, e.g., the runway. The pressure, upon which the relative altitude is determined, is specified as ***Question Field Elevation (QFE)***.

While setting this pressure, the altimeter will display the real aircraft's flight height toward the airfield runway. If the aircraft is on the runway itself, the altimeter will show zero. Since the airfields' runways are situated on different altitudes toward the sea level, the pressure on these runways will be different at the same atmosphere pressure.

Pressure altitude H_{pa} (called ***Relative Barometric Altitude*** in Russia) is the altitude, measured toward the level which is equal to the atmosphere pressure 1013.2 HPa (760 mm of Mercury). Relative barometric altitude is also the ***Flight Level (FL)***, because exactly on it the vertical separation of airspace on the routes and in the waiting areas is set. The pressure, at which the altitude of FL is defined, is marked as ***Question Nautical Elevation (QNE)***.

While setting the QNE pressure, the altitude displayed by the altimeter will not be pegged neither to the real flight altitude nor to the changeable atmosphere pressure. The purpose of setting the standard pressure is that the flight crew will not have to set the pressure every time the weather area is changed. Thereat, all aircrafts that set

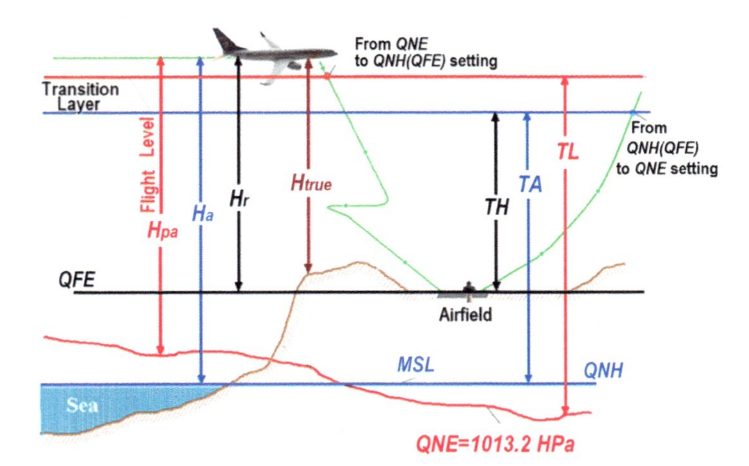

Fig. 1.5 To the flight altitude and height determination

the standard QNE pressure will have single altitude reference point, thus during the flight on various altitudes secure vertical interval will be provided.

Rules of flights procedure include the conversion of the barometrical altimeter reference scale when changing the echelon height H_r to absolute height H_a (***TA—Transition Altitude***) or H_r (***TH—Transition Height***) when landing and vice versa when taking off. These operations are held in the ***Transition Layer (TL)***—airspace between the conversion's height and echelon. Aircrafts' flights in the horizontal flight mode are prohibited in the transition layer.

Track is the projection of a flight trajectory to the Earth surface. ***Desired track***, along which the aircraft's center of mass must move and the ***true track***, along which it moves under the disturbing factors (wind, navigation parameters' measuring errors, piloting errors, etc.) is distinguished (Fig. 1.6).

Aircraft's ***heading*** is the angle in the horizontal plane between the direction, taken for the reference point, the aircraft's center of mass passing through, and its roll axis. Thereat, it is assumed in navigation that all angles are reckoned clockwise from the direction, taken for the reference point.

Depending on the direction, taken for the reference point, true and magnetic headings are distinguished (Fig. 1.7).

True heading $\boldsymbol{\Psi}_T$ is reckoned from the grid north of true meridian line N_T, ***magnetic heading*** $\boldsymbol{\Psi}_M$—from the grid north of magnetic meridian line N_M.

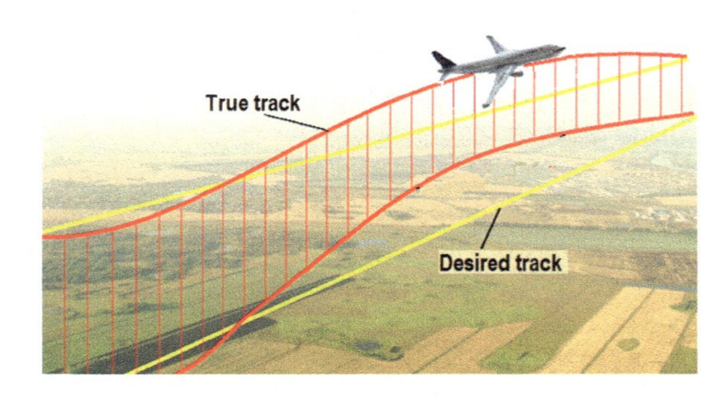

Fig. 1.6 Desired and true tracks' lines

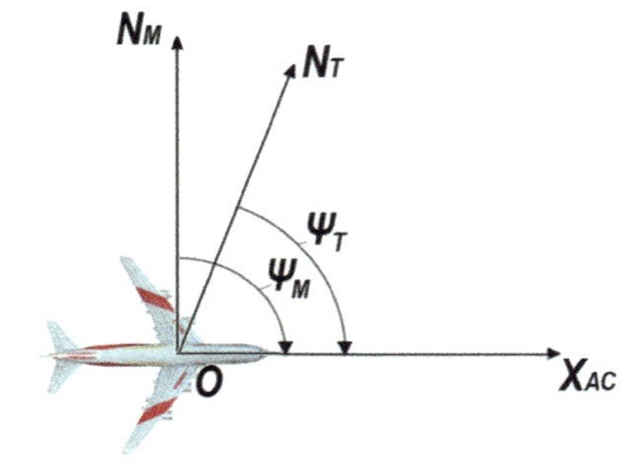

Fig. 1.7 True and magnetic headings

Relative bearing $\boldsymbol{\Theta_R}$ is the angle in horizontal plane between the roll axis of an aircraft, taken for the reference point, and the direction to the radio landmark or radio beacon (Fig. 1.8).

True bearing $\boldsymbol{\Theta_T}$ is the angle in horizontal plane between northbound of true meridian, passing through the aircraft's center of mass, and the direction to the radio landmark. It is often that instead of using the true bearing concept the ***azimuth*** one is used.

Magnetic bearing $\boldsymbol{\Theta_M}$ of a radio beacon, ***true bearing*** $\boldsymbol{\Theta_{AC}}$ of an aircraft, and ***magnetic bearing*** of an aircraft are distinguished (Fig. 1.8).

Meridians, passing through the aircraft's center of mass and the radio landmark when projecting on the surface, are not parallel (N_T and N_T') since all meridians converge in the Earth poles. That's why for Fig. 1.8 the following equation is relevant:

$$\Theta_T = \Theta_{AC} \pm 180° \pm \Delta,$$

where Δ—is the angle of meridians' convergence.

Air, ground, and vertical speed are distinguished when describing the aircraft's speed.

Air speed $\vec{V}$ is the horizontal component of an aircraft's center-of-mass motion speed toward the atmosphere. Air speed vector directs to the roll axis of an aircraft.

Ground speed $\vec{W}$ is the horizontal component of an aircraft's center-of-mass motion speed toward the Earth surface.

Ground speed vector is equal to the sum of air speed and wind speed $\vec{U}$ vectors. These vectors make so-called navigation triangle of speeds (Fig. 1.9).

The angle between air and ground speeds' vectors is called the ***drift angle (DA)***. The drift angle is positive if the ground speed's vector is directed right toward the roll axis of an aircraft.

Actual track angle (ATA) is the angle between the grid north, passing through the aircraft's center of mass and the ground speed vector.

Wind angle (WA) is the angle between the grid north, passing through the end of an air speed vector and the wind vector.

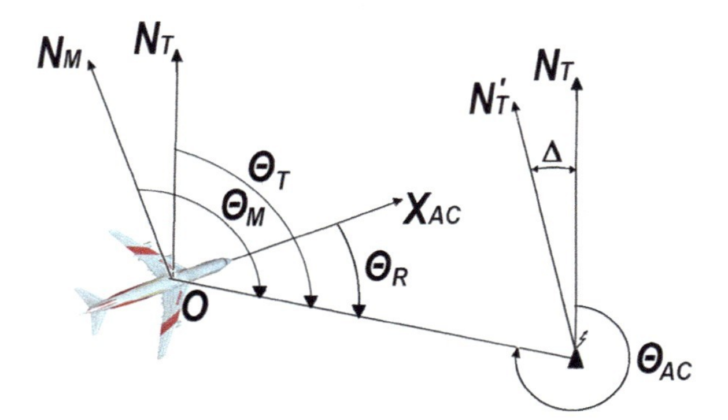

Fig. 1.8 Relative bearing and absolute (true and magnetic) bearing

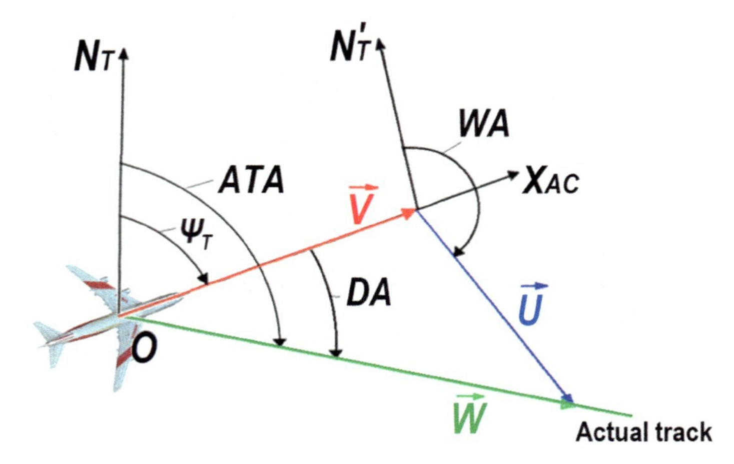

Fig. 1.9 Navigation triangle of speeds

Vertical speed is the speed of an aircraft's center-of-mass movement in the vertical plane.

Roll, pitch, and yaw describe the rotary motions of an aircraft toward its center.

Pitch angle is the angle between the roll axis of an aircraft and the local horizontal plane. In aviation, pitch with the angle increase (pitch-up) and angle decrease (pitch-down) is distinguished.

Roll angle is the angle between the aircraft symmetry plane and horizontal plane that describe the turn of an aircraft around the roll axis.

Yaw angle describes the aircraft position when turning around the vertical axis.

1.4 Coordinate Systems Used in Air Navigation

For the aircraft's position and motion parameters determination as well as the mathematical description of navigation processes, a reference system should be specified. This is called the coordinate system.

The chosen coordinate system must meet the number of requirements:

navigation tasks solving with the required accuracy;
coverage of necessary Earth surface territory square or the airspace capacity, in the range of which the navigation tasks are solved;
visibility and simplicity of information display and perception regarding the object's position in the axis system;
getting the simplex mathematical correlations that describe the aircraft's motion process.

Requirements listed above are controversial. Thus, the coordinate systems' choice, being the same for the whole earth surface, inevitably brings to complicated mathematical correlations, and the coordinate systems that allow to solve navigation tasks using relatively simple mathematic dependencies provide the acceptable accuracy only in limited area of space. That is why in practice different coordinate systems can be used, each of which provides the most effective solution of specific navigation tasks.

Coordinate systems used in air navigation can be classified by the following features:

by the amount of Earth's surface or spatial region coverage (local, global);
by the position of coordinate system's origin (geocentric—the origin matches the Earth center of mass, topocentric—the origin is in the point on the Earth surface and also the coordinate systems connected with an aircraft and moving together with it toward the Earth surface);
by plane's reference orientation (horizontal, equatorial, orbital).

The inertial coordinate system that is fixed in the universal space needs to be marked out. This coordinate system is perfect for aircrafts' and space modules' navigation, since the classical mechanics' laws are strictly obeyed in it.

Local coordinate systems cover limited area of Earth surface and are used at aircraft's movements to the distances up to 400–450 km, when the Earth curvature can be neglected for navigation task solution's accuracy. In such coordinate systems, landing systems (ILS) and short-range navigation systems VOR/DME, TACAN, are used.

Local coordinate systems include ***cylindrical, spherical, and orthogonal Cartesian*** (Fig. 1.10) which origins are situated in a point on the earth surface. These are topocentric systems.

In these coordinate systems, *M* point coordinates are:

in Cartesian system (Fig. 1.10a)—x, y, z coordinates;
in cylindrical (Fig. 1.10b)—r projection onto the horizontal plane of radius-vector ρ, traced from the coordinate system's origin O point to M point, azimuth Θ, height z;

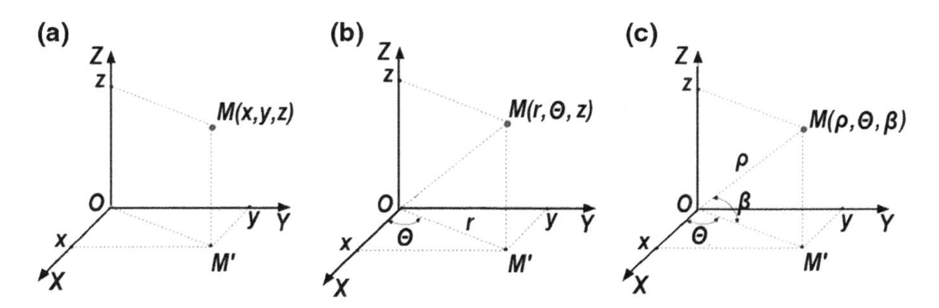

Fig. 1.10 Local coordinate systems

in spherical (Fig. 1.10c)—distance ρ (radius-vector) to the M point from the coordinate system's origin, azimuth Θ, elevation β. Spherical coordinate system is also called "rho-theta" system.

Azimuth is reckoned from the *OX*-axis direction in horizontal plane, which as a rule is oriented to the magnetic north direction, till the radius-vector r projection onto the horizontal plane. *OZ*-axis is oriented to Earth surface in *O* point on the normal.

Local coordinate systems are widely used in short-range flights (up to 500 km), aircraft control on all taking off, approach, and landing phases, and aircraft position determination toward the landmarks.

There are pretty elementary mathematic correlations (transformational coordinates) that allow to re-count the coordinates of a point from one coordinate system to another.

Transformation (1.1) from the cylindrical to Cartesian system and reversed transformation (1.2) is written as:

$$\begin{aligned} x &= r\cos\Theta, \\ y &= r\sin\Theta, \\ z &= z. \end{aligned} \tag{1.1}$$

$$\begin{aligned} r &= \sqrt{x^2+y^2}, \\ \Theta &= \operatorname{arctg}\frac{y}{x}, \\ z &= z. \end{aligned} \tag{1.2}$$

Transformation (1.3) from spherical into the Cartesian system and reversed transformation (1.4) is written as:

$$\begin{aligned} x &= \rho\cos\beta\cos\Theta, \\ y &= \rho\cos\beta\sin\Theta, \\ z &= \rho\sin\beta. \end{aligned} \tag{1.3}$$

$$\begin{aligned} \rho &= \sqrt{x^2+y^2+z^2}, \\ \Theta &= \operatorname{arctg}\frac{y}{x}, \\ \beta &= \operatorname{arctg}\frac{z}{\sqrt{x^2+y^2}}. \end{aligned} \tag{1.4}$$

The advantages of described coordinate systems include pretty elementary mathematical correlations that describe aircrafts' motion processes, visibility, and simplicity of display and perception of information about object's position toward the origin of coordinate system, simple and accurate coordinate transformations from one system to another. Main disadvantage of these systems includes small fraction of Earth surface's parts coverage.

Global coordinate systems cover the whole Earth's surface. Earth's figure and therefore its surface has an irregular shape (Fig. 1.11a). Unfortunately, there is no such coordinate system that considers the Earth's figure absolutely accurate while describing the navigation processes toward the earth surface in any of its area. That is why various approximations of Earth's figure are used for meeting the accuracy requirements when solving geodesic, cartographic, and navigation tasks.

The usage of modern technologies in Earth's parameters measurements, satellite navigation development, and also the requirements of airspace interoperability caused significant changes in approach to describing the Earth's figure and this description accuracy. As a result, today such coordinate systems are used for geodesic and cartographic, as well as for air and cosmic navigation tasks solving.

Systems of the first type that are used for quite a long period of time are oriented on the separated determination of objects' position on the Earth surface (horizontal 2D space) and vertically (orthometrical altitude which is reckoned from the mid sea level (MSL)) and systems of the second type—on the objects' position determination in the 3D space. In both cases, more accurate approximation of Earth's figure and its surface is required.

Geoid is the closest in shape to the Earth's surface. Geoid (Fig. 1.11b) is the geopotential surface of the Earth's gravity field, matching the MSL's surface in its calm state. Due to effects, such as atmospheric pressure, temperature, prevailing winds and currents, and salinity variations, MSL will depart from this level surface by a meter or more.

Although the geoid surface is flat in comparison to the physical surface of Earth, it still has irregular shape. This is caused by erratic positions of gravitational masses in the Earth's body, causing the plumb-lines deviation. Geoid is the surface toward which the heights reference to MSL is being reckoned from.

To create the global geoid, the Earth Gravitational Model was developed in 1996—EGM96 and geoid WGS-84 (EGM96) passed that provides the accuracy of as well as 1 m in the points where gravitation was measured. The WGS-84 Earth Gravitational Model 2008 (EGM2008) is the latest, most accurate, and complete gravitational model from which a global geoid is derived. This supersedes EGM96 which is the previous model.

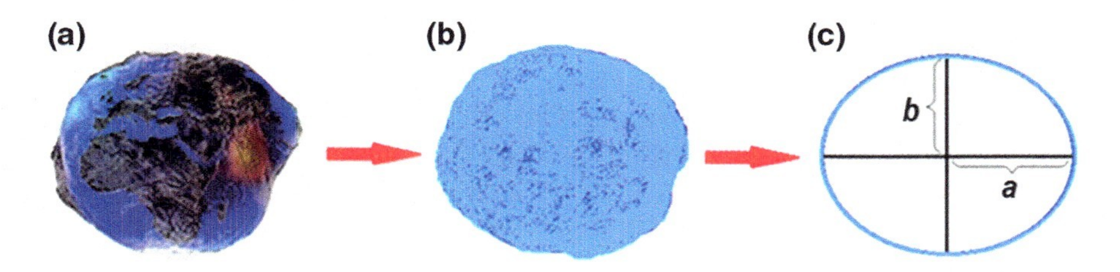

Fig. 1.11 Earth figure approximations

Geoid cannot be mathematically described, that is why for the practical tasks solving on the Earth's surface, it is showed up as mathematically described geometric figure—ellipsoid (Fig. 1.11c).

Any rotating liquid homogenous body of big mass can take ellipsoid shape under inner gravity and centrifugal forces. By picking the ellipsoid parameters, it can be more or less moved closer to the geoid in its different parts.

This may mean that the geoid does not fit the actual Earth in another part of the world. The differences in heights referenced to the geoid versus heights referenced to the ellipsoid can be as much as 100 m.

Till 1964, every country selected the parameters of an ellipsoid that was close to the geoid within the territory of that country. Such ellipsoid was named the ***local ellipsoid***.

Normally, local ellipsoids were taken for legislatively geodesic measurements' processing. Historically, different ellipsoids were taken and legislated at different periods of time in various countries and their parameters didn't coincide.

In Russia/USSR, the ***Krasovsky ellipsoid*** was used with the following parameters: semi-major axis $a = 6{,}378{,}245$ m, semi-minor axis $b = 6{,}356{,}863$ m, compression ratio 1:298.3. In the USA and Canada, the ***Clarke ellipsoid*** (Clarke 1880) was used with the following parameters: semi-major axis $a = 6{,}378{,}249$ m, compression ratio 1:295.0. In many western Europe and some Asian countries, ***Hayford ellipsoid*** was taken, while in former British colonies such as India and Southern Asia counties, the ***Everest ellipsoid*** was used.

Aircrafts' flight distances increase, global navigation satellite systems' introduction as well as the flight management systems' (FMS) development detected the number of critical problems when using local ellipsoids, e.g., measured coordinates' kicks when passing different countries' airspace. That is why the work has been carried out to set the global ellipsoid which would be suitable for all countries. As a result, the geodesic reference system 1980 (GRS-80) ellipsoid was adopted.

"Geodesic positions referenced to the Earth are defined in the general context of a ***terrestrial reference system*** and with respect to a specific ***terrestrial reference frame***. The reference system defines the physical constants (gravitational constant, the semi-major axis of the Earth's best fitting ellipsoid, the speed of light), models and coordinate system needed to unambiguously and consistently define the coordinates of a point.

As an example, the coordinate system can be defined as a three-dimensional Cartesian (*OXYZ*) system with its origin *O* point at the Earth's center of mass and the three coordinate axes aligned with the equator and the rotational axis of the Earth, and rotating with the Earth's crust" [1].

"The scientific standard for the terrestrial reference system is the International Terrestrial Reference System (ITRS). The ITRS embodies a set of conventions that represent the state of the art for referencing geodesic positions to the Earth. These conventions are established by the International Earth Rotation and Reference Systems Service (IERS)" [1].

In ITRS, it is convenient to describe the navigation processes using the Inertial Reference Systems (IRS) and satellite navigation systems, since the measurements in the first case are performed in inertial space, in the second case—toward the objects, which are not connected with the earth's surface.

The physical realization of this system represents a global network of ground stations (on the Earth's crust) whose three-dimensional coordinates and linear velocities are derived from space-based observations. This station set defines the International Terrestrial Reference Frame (ITRF). The ITRF is periodically modified and applies any changes that have been adopted in the ITRS.

The current version of the reference frame is ITRF-2008.

The USA's terrestrial reference system is WGS-84. WGS-84 constitutes an Earth-centered and Earth-fixed coordinate system and a prescribed set of constants, models, and conventions that are largely adopted from the ITRS. Its semi-major axis $a = 6{,}378{,}137$ m, semi-minor axis $b = 6{,}356{,}777$ m, compression ratio 1:298.257.

ICAO has adopted WGS-84 as the world standard and designated the use of the WGS-84 as the universal datum.

In the USSR, the state terrestrial reference system "Parametry Zemli (Earths parameters) 1990" (PZ-90) was introduced. In November 2007, the system was modified and named PZ-90.02. Its parameters were modified for some meters at once, and it started to match WGS-84 (semi-major axis $a = 6{,}378{,}136$ m, semi-minor axis $b = 6{,}356{,}777$ m, compression ratio 1:298.258).

The current version of the Russia terrestrial reference system is PZ-90.11, set in 2012. Ellipsoid's PZ-90.11 surface is taken as a vertical reference in this coordinate system.

"In the USA, the North American Datum 1983 (NAD-83) is the standard geodetic reference system that defines three-dimensional control for the country. The GRS-80 ellipsoid was adopted as the reference surface. Ellipsoid heights are also associated with the traditional horizontal control points to define a rigorous set of 3D coordinates" [1].

In Russia, SK-2011 (Sistema Koordinat-2011) is the current version of standard geodesic reference system since 2012. The GRS-80 ellipsoid was adopted as the reference surface for it. Its previous versions are SK-42 and SK-95, in which the Krasovsky ellipsoid was used as the reference surface.

NAD-83 and SK-2011 are made for performing geodesic and cartographic works. These systems are convenient for aircraft's navigation toward the Earth stations and landmarks, which coordinates are tied to the Earth's surface.

Vertical data are traditionally associated with MSL. All elevations on land are referenced to this zero value.

"Many national and regional vertical data are tied to a local mean sea level (LMSL), which may significantly differ from global MSL due to local effects such as river runoff and extremes in coastal tidal effects. Thus, national and regional vertical data around the world, which are tied to LMSL, will significantly differ from one another when considered on a global basis. In addition, due to the ways the various vertical data are realized, other departures at the meter level or more will be found when comparing elevations to a global geoid reference" [1].

In the USA, the North American Vertical Datum 1988 (NAVD 88) is used for the altitude reference (the single tide gauge elevation at Point Rimouski, Quebec,

Canada, as the continental elevation reference point and essentially references all other elevations in the USA to this).

In Russia, all altitudes take their origin from the Baltic Sea level (from the zero-point of flagstaff located in Kronstadt, Saint-Petersburg). Today, the Baltic Sea heights' system of 1977 is used.

In order to express coordinates in geodetic terms as longitude B, latitude L and ellipsoid height H, a two-parameter reference ellipsoid is defined. For geocentric terrestrial reference systems (Fig. 1.12), this ellipsoid is selected so that its center (O point) coincides with the Earth's center of mass, its axes are oriented and fixed to the ITRS coordinate axes, and its semi-major a and semi-minor b axes and rotation rate approximate those of the Earth. The semi-major axis of the ellipsoid coincides with the OZ-axis of the ITRF, while the OX- and OY-axes of the ITRF are fixed to the ellipsoid on its equatorial plane. The OZ-axis is rotating at a rate that approximates that of the Earth.

Geodesic (or Geographic) latitude of M point is the angle between the normal line to the ellipsoid's surface in this point and the equator's surface (Fig. 1.13a, c). This normal line does not necessarily pass through the center of the ellipsoid.

The latitude is reckoned from the equator's plane to poles from 0 to $\pm 90°$ (north and south latitude).

Geodesic (or Geographic) longitude of M point is the dihedral angle between the Greenwich (Prime) meridian planes and the meridian passing through the M point (Fig. 1.13b, c).

The longitude is measured on either side of Greenwich meridian in the range from $0°$ to $\pm 180°$ (east or west longitude). The meridian passing through Greenwich (in Britain) is known as the ***prime meridian or international reference meridian***.

Geodesic (orthometric) altitude H of M point is the distance on the normal from the ellipsoid's surface to the M point.

Ellipsoid's surface has strict mathematical description and allows to get the formulas for navigation tasks solving with high accuracy. However, these formulas are pretty complicated and are realized in practice only in modern FMS's computers.

Connection between the object's coordinates in Cartesian system $OXYZ$ and geodesic (spatial ellipsoidal) coordinates is determined by the following equations:

$$x = (N + H) \cos B \cos L$$

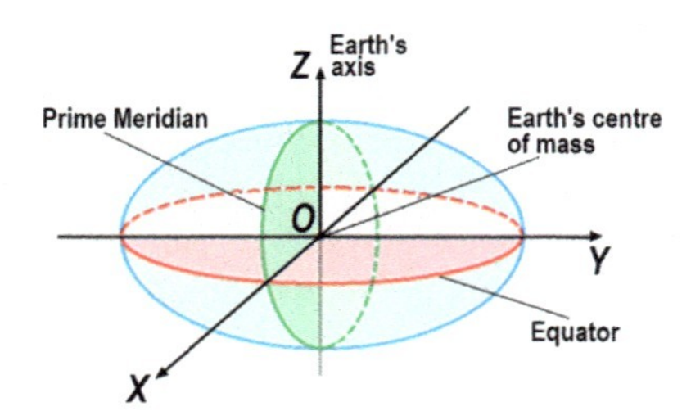

Fig. 1.12 Geocentric terrestrial reference systems

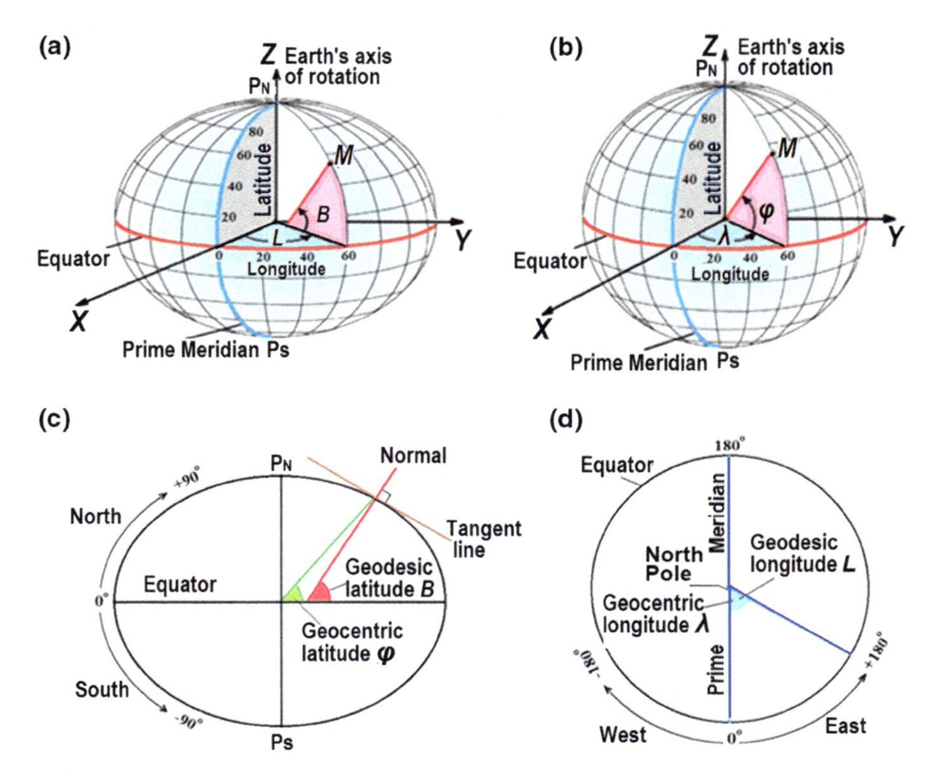

Fig. 1.13 Global geocentric coordinate systems

$$y = (N + H)\cos B \sin L$$
$$z = (N + H - e^2 N)\sin B,$$

where H is the height above ellipsoid's surface,

$e = \frac{\sqrt{a^2-b^2}}{a}$—eccentricity of a ground ellipsoid,
$N = a(1 - e^2 \sin^2 B)^{-1/2}$—radius of the first vertical's curvature.

Coordinates' transformation from Cartesian system *OXYZ* into geodesic (spatial ellipsoidal) is performed using the following equations:

$$B = \mathrm{arctg}\frac{z}{\sqrt{(x^2+y^2)}}\left(1 - e^2\frac{N}{N+H}\right)^{-1}$$
$$L = \mathrm{Arctg}\frac{y}{x}, \quad [-\pi, \pi]$$
$$H = \frac{\sqrt{x^2+y^2}}{\cos B} - N,$$

where Arctg(*) is a circular value of an anti-tangent considering the quadrants.

It is important to know that this transformation is performed using the iterational method and the convergence is achieved fast.

For the objects' coordinates translation into the local topocentrical coordinate system ($OX_{\text{Loc}}Y_{\text{Loc}}Z_{\text{Loc}}$), being geodetically connected to the predefined Earth's point (B_0, L_0, H_0) (Fig. 1.14), first the translation is performed from geodesic coordinates (B, L, H) into the geocentric rectangular coordinates (x, y, z), connected to the Earth's center, and then the transformation is performed as:

$$\begin{vmatrix} x_{\text{Loc}} \\ y_{\text{Loc}} \\ z_{\text{Loc}} \end{vmatrix} = \boldsymbol{M} \begin{vmatrix} x - x_0 \\ y - y_0 \\ z - z_0 \end{vmatrix},$$

where x_0, y_0, z_0 are geocentrically orthogonal coordinates of the local coordinates system's origin,

$$\boldsymbol{M} = \begin{vmatrix} -\sin L_0 & \cos L_0 & 0 \\ -\cos L_0 \sin B_0 & -\sin L_0 \sin B_0 & \cos B_0 \\ \cos L_0 \cos B_0 & \sin L_0 \cos B_0 & \sin B_0 \end{vmatrix} - \text{transformation matrix.}$$

Reversed translation of local coordinates ($OX_{\text{Loc}}Y_{\text{Loc}}Z_{\text{Loc}}$) into the geocentrical coordinate system ($OXYZ$) is performed according to the equation:

$$\begin{vmatrix} x \\ y \\ z \end{vmatrix} = \boldsymbol{M}^{\text{T}} \begin{vmatrix} x_{\text{Loc}} \\ y_{\text{Loc}} \\ z_{\text{Loc}} \end{vmatrix} + \begin{vmatrix} x_0 \\ y_0 \\ z_0 \end{vmatrix},$$

where $\boldsymbol{M}^{\text{T}}$—is the transponded matrix.

For simplifying the mathematical description of navigation processes, ellipsoid is replaced by the sphere and ***geospherical coordinate system*** is applied (Fig. 1.13b).

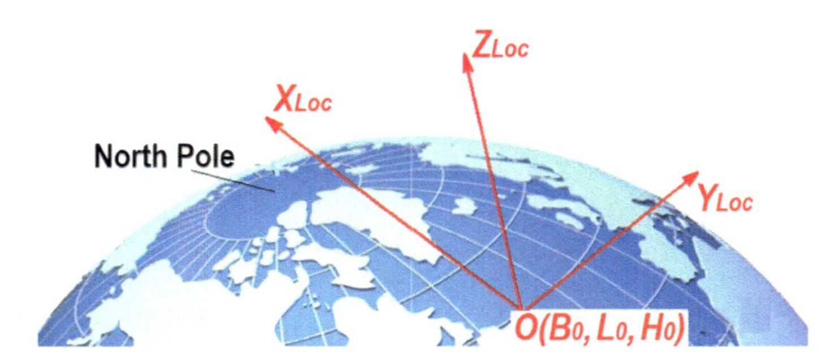

Fig. 1.14 Topocentrical local Cartesian system

The M point position on its surface is determined by geospherical coordinates—latitude φ and longitude λ.

Yet, the geospherical (geocentric) latitude of M point is the angle between the radius-vector, lined from the spheric center to the M point, and the equatorial plane.

Concepts and the way of geospherical λ and geodesic longitude L reckoning match.

When solving the navigation tasks in geospherical coordinate system, methodical errors of determining the object's position appear which are caused by the approximation of Earth figure by the sphere. The maximum difference between geocentric and geodesic latitudes is most significant in mid-latitudes (at approx. 45°N/S) and makes $(B - \varphi)_{\max} = 11.6$ min of arc and decreases while approaching the poles or equator. Upon that, geospherical latitude is always less than geodesic one (except for the poles and equator) and geodesic and geospherical longitudes' points match.

The geodesic latitude re-counting into the geospherical one can be performed using the approximate equation:

$$\varphi = B - 8'39'' \sin 2B.$$

For the point being on the Earth's sphere surface, transformations from geospherical coordinate system into the Cartesian system $OXYZ$ and vice versa are written as

$$\begin{aligned} x &= R \cos\varphi \cos\lambda, \\ y &= R \cos\varphi \sin\lambda, \\ z &= R \sin\varphi, \end{aligned}$$

$$\begin{aligned} R &= \sqrt{x^2 + y^2 + z^2}, \\ \varphi &= \operatorname{arctg}\frac{z}{\sqrt{x^2 + y^2}}, \\ \lambda &= \operatorname{arctg}\frac{y}{x}. \end{aligned}$$

For navigation parameters measurement with the help of onboard technical aids which antennas system directional patterns are oriented to the construction axes of an aircraft, aircraft-related coordinate systems are used. Cartesian system $OX_{ac}Y_{ac}Z_{ac}$ is normally used, which origin is combined with the aircraft's center of mass, OX_{ac}-axis matches its longitudinal axe, OZ_{ac}-axis is directed to the right semi-plane side, and OY_{ac}-axis is directed upward. Such coordinate system is moving in space according to aircraft's evolutions.

For solving the navigation tasks toward the Earth surface, this method turns out to be inconvenient in the number of cases, so the horizontal coordinate system $O_{\text{hor}}X_{\text{hor}}Y_{\text{hor}}Z_{\text{hor}}$ is used (Fig. 1.15) that is connected to an aircraft where the $X_{\text{hor}}O_{\text{hor}}Z_{\text{hor}}$ surface keeps horizontal position at aircraft's evolutions.

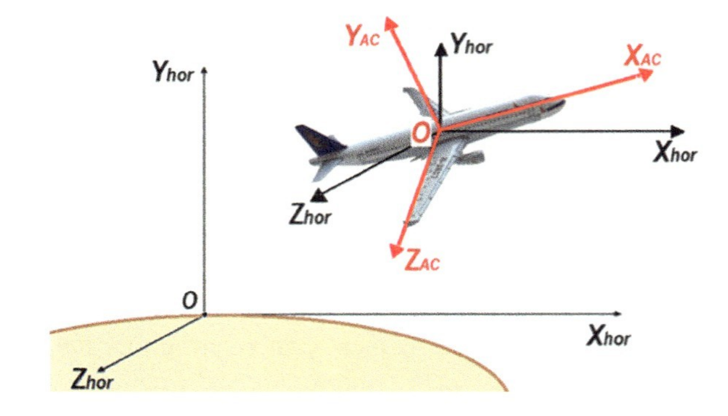

Fig. 1.15 Aircraft-related coordinate systems

For re-counting of measured navigation parameters from the aircraft's into horizontal coordinate system, the information about aircraft's roll and pitch is required.

1.5 Lines and Surfaces of Position

Navigation parameters' measurements such as the distance, range difference, true bearing, that are carried out in radio navigation systems, allow to determine the ***lines of position*** (LOP) and ***surface of position*** (SOF).

Line (surface) of position is the geometric locus on a plane (in space), where each point has the constant navigation parameter value.

It should be highlighted that this definition cannot fit to the widely used line of position in navigation, which is called the ***Great Circle*** (***Orthodromic Route***).

A circle on the surface of the Earth which center and radius are those of the Earth itself is called a Great Circle. It is called "great" because a disk cut through the Earth in the plane of the Great Circle would have the largest area that can be achieved. Great Circle makes sense only in such cases, when the Earth surface is approximated by the sphere.

Great Circle route is the shortest distance between two points on the Earth's surface (is the shorter arc of the Great Circle joining the two points) (Fig. 1.16a).

Equator is the Great Circle, which surface is perpendicular to the Earth's roll axis.

It is important to note that for the geodesic coordinates' system (while approximating the Earth's surface by the ellipsoid), the concept of Great Circle is not applicable. The shortest distance line between two points on the ellipsoid's surface is called the geodesic line.

The Great Circle (in km) distance between two points on the Earth's surface is defined by the following equation:

$$S_{GC} = 111.12 \arccos(\sin\varphi_1 \sin\varphi_2 + \cos\varphi_1 \cos\varphi_2 \cos(\lambda_2 - \lambda_1)),$$

where (φ_1, λ_1), (φ_2, λ_2)—coordinates of orthodromic points; arccos(*) value is expressed in degrees.

Rhumb Line (***Loxodromic Line***) (Fig. 1.16b)—a line which cuts or crosses all meridians at the same angle α_{RL}. Rhumb line is the helix on the sphere's surface, which is asymptotically moving closer to the pole. The flight on the Rhumb Line is conducted at the constant true heading (TH = const) with the lack of drift angle.

Distance between two points located on Rhumb Line is defined by the following equation:

$$S_{\mathrm{RL}} = 1.85(\varphi_1 - \varphi_2)/\cos\alpha_{\mathrm{RL}},$$

where α_{RL}—the course angle of Rhumb Line which is being calculated based on the equation:

$$tg\alpha_{\mathrm{RL}} = \cos\varphi_{im}(\lambda_2 - \lambda_1)/(\varphi_2 - \varphi_1),$$

where $\varphi_{im} = (\varphi_1 + \varphi_2)/2$—mid-latitude of Rhumb Line; 1.85–1 min line value of meridian arc; (φ_1, λ_1), (φ_2, λ_2)—Rhumb Line points' coordinates.

The distance on the Rhumb Line is bigger than on the Great Circle route, except for the alongside meridian or equator's flight cases.

The ***Curve of the Equal Bearings*** (CEBs) (Fig. 1.17) is the line of position, in every point of which the direction on the specified radio beacon (point *B*) makes the constant angle α_0 with the meridian.

This curve is the line of aircraft position, which is conducting a flight with constant true radio bearing (TRB = const).

Curve of the Equal Angles (CEAs) (Fig. 1.18) is the line of position, in every point of which the angle β between directions on two radio beacons (RB1 and RB2) is the constant value. On the surface, CEA is the arc of circle, traced through the chosen RB in such a way that the angle β matches the set value.

The flight on CEA can be conducted by determining relative bearing Θ_1 and Θ_2 of two radio beacons by fulfilling the condition

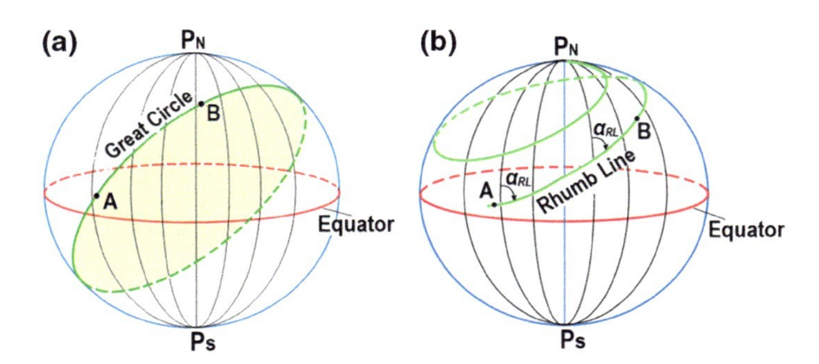

Fig. 1.16 Great Circle route and Rhumb Line on the sphere's surface

$$\beta = \Theta_1 - \Theta_2 = \text{const.}$$

Circle of position (COP) is the line of position, where each point is equally spaced from radio beacon RB (Fig. 1.19). COP is a circle on the surface, in the middle of which the RB is situated. For the flight on COP, the condition R = const must be fulfilled.

Surface of position is a sphere, in the middle of which the RB is situated.

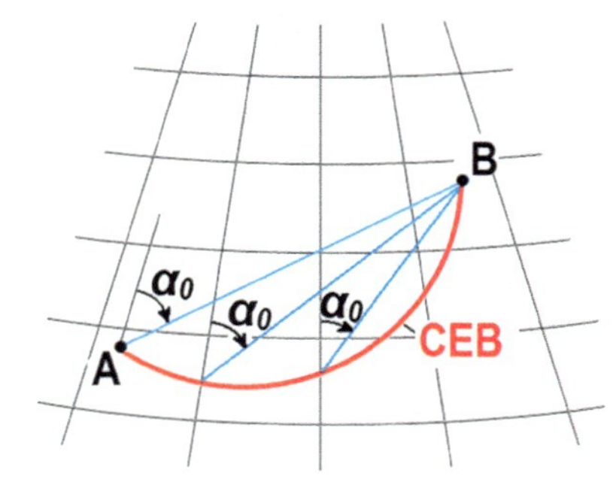

Fig. 1.17 Curve of the equal bearings

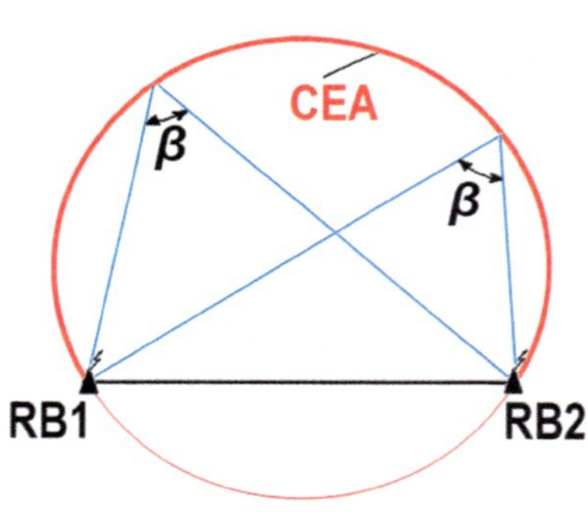

Fig. 1.18 Curve of the equal angles

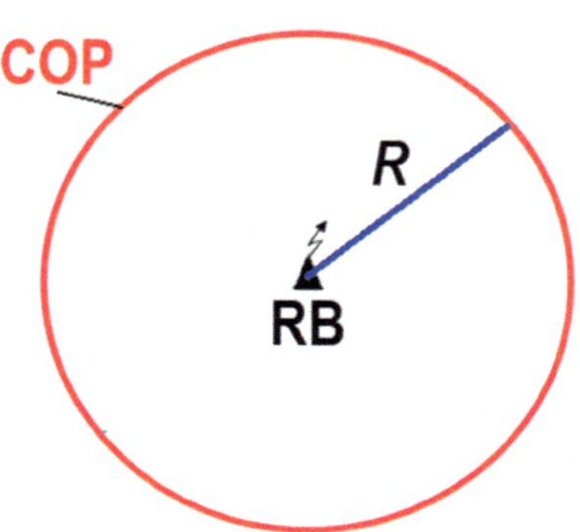

Fig. 1.19 Circle of position

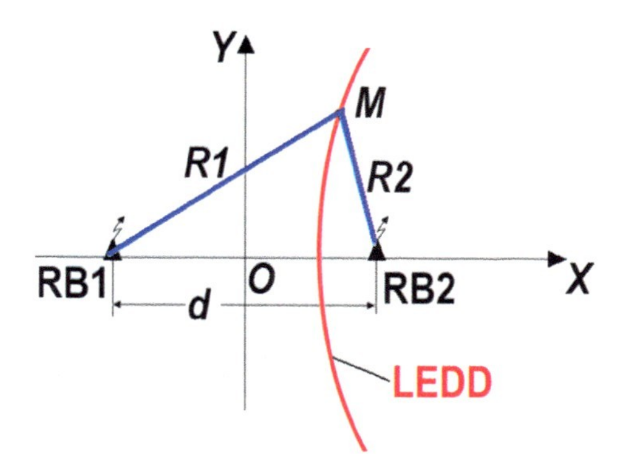

Fig. 1.20 Line of the equal differences in the distance

Line of the equal differences in the distance (LEDD) is the line of position, in each point of which the range difference to two RB is constant (Fig. 1.20). On the surface, LEDD is a hyperbole, in which focus the RB is situated.

For the flight on LEDD, further condition must be fulfilled:

$$|R_1 - R_2| = \text{const.}$$

Circle of position is the line of position at navigation determinations with the help of distance-measuring (rho-rho) radio navigation systems. Line of the equal differences in the distance is the line of position in differential distance-measuring (hyperbola fixing) radio navigation systems.

Therefore, the measurement of a single navigation parameter gives a single line of position on the surface. This is enough for setting and following the route of flight along the position line. However, for determining the aircraft's position at least two intersecting lines of position are required, i.e., measure two navigation parameters.

1.6 Methods of Aircraft's Position Determination

For determining the aircraft's position, three methods are used: ground-mapping and matching, dead reckoning, and intercept or LOP method.

Ground-mapping and matching methods are visual orienting, comparing of radio-locating, optical or television image of terrain which is received on the board of an aircraft with the relevant maps, correlation-extremal navigation on the physical fields of Earth.

In the simplest case, the ground-mapping and matching method of aircraft's position determination is based on comparison of the terrain's image on the map or on the navigation display with the Earth's surface view, which is being observed by the crew visually or with the help of technical aids (e.g., ground-mapping radar) (Fig. 1.21).

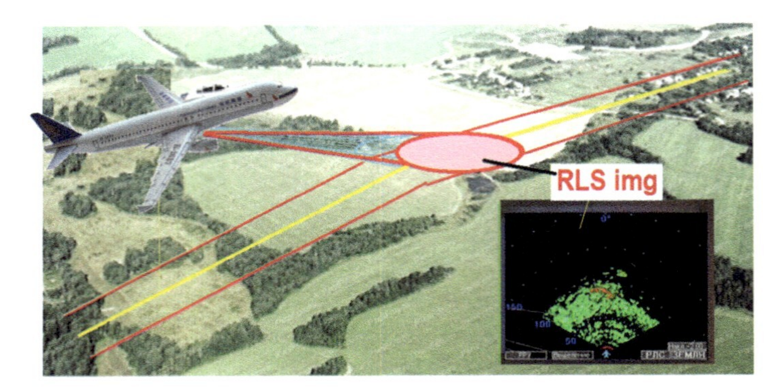

Fig. 1.21 Ground-mapping and matching method

In practice, the error rate of aircraft's position determination using this method is 1–3 km when away the 5–15 km from beacons and 0.1–0.3 km when passing above the beacon. Whereas the navigation beacons are natural or man-made, well seen on the common landscape objects (inhabited area, detached buildings, bridges, etc.) with certain coordinates. This method is pretty simple but it has low accuracy that worsens with height increase and flight speed. The disadvantage of this method is the inability of determining the position in bad weather conditions, at night, as well as while flying over the featureless terrain (sea, forest, steep, etc.).

Correlation-extremal navigation is based on some physical field's parameters determination which is typical for that terrain and its comparison (by calculating the cross-correlation function) with parameters, inserted into the memory storage of onboard correlation-extremal system for the designated flight route. The information about the terrain's relief height field, gravitational or magnetic field along the flight route can be used.

This method has quite high potential accuracy, however for its realization a big volume of a priori information about the parameters of physical field being used on the whole flight route that requires a big volume of calculating system's memory as well. Besides, the calculation of cross-correlation function and flight path control signals' output in real time represents quite a complicated task. As to describe reasons, correlation-extremal systems nowadays have not found its use in aircraft navigation.

Dead reckoning method is based on the aircraft's coordinates determination by integrating velocity components timing-wise into the flight management and guidance computer (FMGC) or by double integration of measured accelerations. The simplest dead reckoning systems measure aircraft's heading and speed, resolve speed into the navigation coordinates, and then integrate to obtain position.

Dead reckoning method is used in inertial navigation systems (the most precise dead reckoning system) and in Doppler-monitored heading navigation systems (directional gyroscope, or gyrocompass and Multibeam Doppler radar).

In the simplest case, the equation of dead reckoning while the object is moving along the OX-axis of Cartesian coordinate system looks as:

$$x(t) = x(t_0) + \int_{t_0}^{t} W_x(t)\mathrm{d}t = x(t_0) + \int_{t_0}^{t}\int_{t_0}^{t} a_x(t)\mathrm{d}t\mathrm{d}t, \tag{1.5}$$

where $x(t_0)$—the initial value of preliminarily known object's coordinate; W_x—speed of movement along the x coordinate; a_x—acceleration along the x coordinate; t—time passed from the motion start.

Random errors of velocity (acceleration) measurements contain constant or slowly changing components that have correlation functions of exponential form.

Let us describe the error of velocity measurement as $\delta W_x = \delta W_{0x} + n_x$, where δW_{0x}—constant component of an error, n_x—fluctuational component of additive white Gaussian noise type (AWGN). Then from (1.5), we can get an equation for coordinate x reckoning error of the form:

$$\delta x(t) = \delta x(t_0) + \int_{t_0}^{t} \delta W_x(t)\mathrm{d}t = \delta x(t_0) + \delta W_{x0}(t - t_0). \tag{1.6}$$

Due to (1.6), the reckoning error will increase over time. This effect is the main and distinctive disadvantage of reckoning method, which basically consists in aircraft's position determination accuracy worsening on flight due to the accumulation while integrating slowly changing components of velocity or acceleration measurement errors.

This method has a number of important advantages such as continuity of receiving complete information about parameters of an aircraft's motion (coordinates, velocity components), endurance, and high reliability.

Intercept or LOP method is based on the aircraft's position determination of two or more lines of position crossing (Fig. 1.22) that are received as a result of navigation parameter measurement. These lines of position can be, for example, two lines of fixed bearing's values (azimuth) (Fig. 1.22a), two lines of equivalent distances COP (Fig. 1.22b), a line of equivalent distances COP and a line of fixed bearing value (azimuth) (Fig. 1.22c), two lines of equivalent range differences LEDD (Fig. 1.22d).

LOP method is widely used in radio navigation systems. The advantages of this method are high accuracy which is not dependent on the flight duration, the ability of aircraft's position determination, excluding the distance covered previously. The main disadvantages of this method are the accuracy dependency on the conditions of radio-wave propagation, the aircraft's position toward the ground beacons (on the angle of LOP crossing), non-endurance, the ability of determining the position within the operating zone of radio navigation systems, amenability to radio interference.

Since navigation parameters in radio navigation systems are measured with errors, it leads to the errors in LOP determination.

Given that the errors of LOP determination have random nature, root mean square deviation of measured LOP from its position is used for description. Therefore, the task of aircraft's position determination by using the LOP method has stochastic nature and in practice resolves itself to the determination of spatial region that cir-

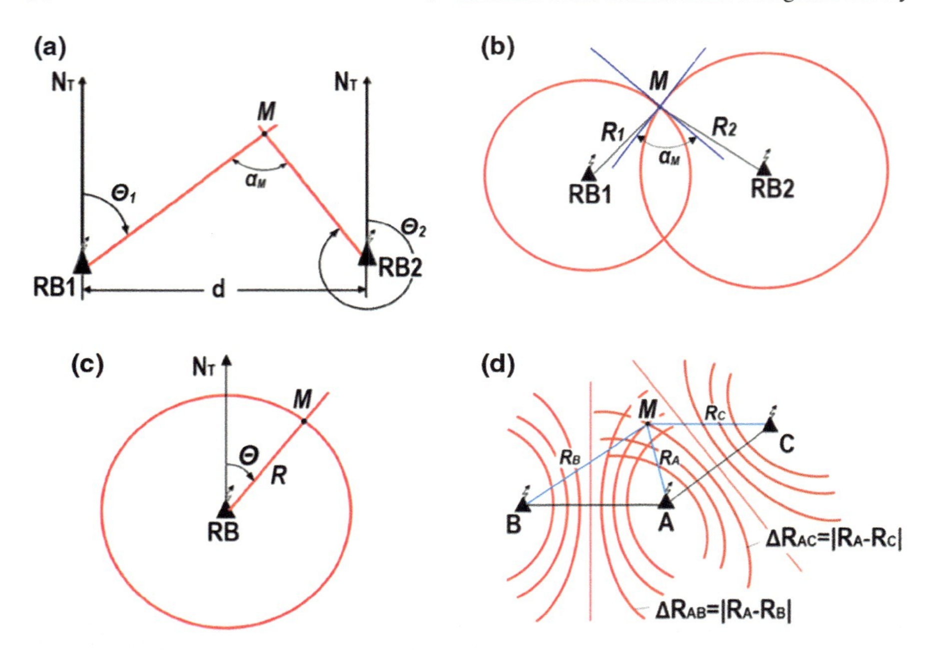

Fig. 1.22 LOP method, **a** theta-theta navigation, **b** rho-rho navigation; **c** rho-theta navigation; **d** hyperbolic fixing navigation

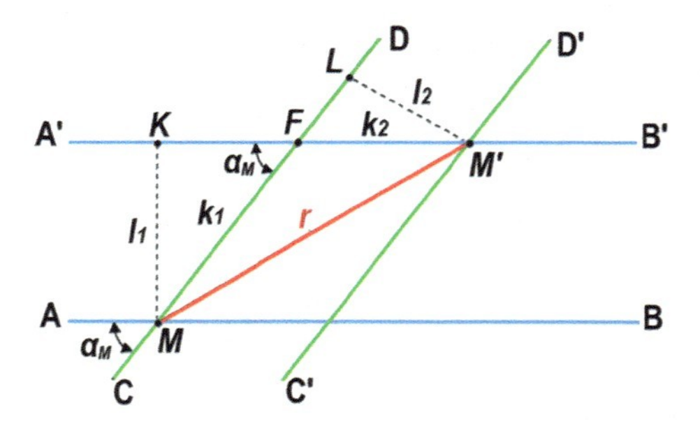

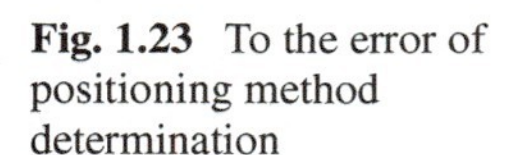
Fig. 1.23 To the error of positioning method determination

cles the true position of an aircraft, inside of which the aircraft's position with the predefined probability determined as a result of measurements is located.

Let us suppose that the aircraft is located on the considerable distance from the ground radio beacons, toward which the LOP is determined. Thus, LOP that match different values of measured navigation parameters in the area of aircraft's position can be replaced by bits of parallel lines regardless of the LOP form (Fig. 1.23).

Let us suppose that the true position of an aircraft is in M point of LOP AB and CD crossing, which coincide the true values of measured navigation parameters. LOP $A'B'$ and $C'D'$ match the aircraft's position which is different from the true one due to measurements' errors. LOP AB and CD are crossed at the angle α_M. Points M and M' are displaced relative to each other at the distance r, which is called *radial error of aircraft's position determination*. Values l_1 and l_2 describe the errors of AB and CD LOP determination.

The task consists in determining the radial error rate r, depending on the errors of LOP determination l_1, l_2, and the angle of LOP α_M crossing. Let us use the Cosine formula for determining this.

Let us mark $k_1 = MF$, $k_2 = M'F$. From the geometrical constructions, it appears that $KM \perp A'B'$, $LM' \perp CD$.

Therefore, we get $k_1 = l_1/\sin\alpha_M$, $k_2 = l_2/\sin\alpha_M$.

From the triangle $MM'F$ by Cosine formula, we see:

$$r^2 = k_1^2 + k_2^2 - 2k_1k_2\cos(180^\circ - \alpha_M) = k_1^2 + k_2^2 + 2k_1k_2\cos\alpha_M.$$

Therefore, the radial error might be expressed as:

$$r = \frac{\sqrt{l_1^2 + l_2^2 + 2l_1l_2\cos\alpha_M}}{\sin\alpha_M}. \tag{1.7}$$

Since the errors l_1 and l_2 are the random values, the radial error of an aircraft's position determination is random. While measuring, it is reputed that expected value (EV) $M\{r\} = 0$, i.e., the systematical error is absent or removed, it leads to the variance. Then for $\alpha_M = \text{const}$, taking into consideration (1.7) we will get the equation for ***radial mean square error*** (RMS) of position determination

$$\sigma_r = \frac{\sqrt{\sigma_{l_1}^2 + \sigma_{l_2}^2 + 2\sigma_{l_1}\sigma_{l_2}\rho\cos\alpha_M}}{\sin\alpha_M},$$

where ρ—coefficient of mutual correlation (mutual dependency) of LOP measurements' errors; σ_l—*mean squared error* (MSE) of LOP determination.

At independent errors of LOP measurements $\rho = 0$ and then:

$$\sigma_r = \frac{\sqrt{\sigma_{l_1}^2 + \sigma_{l_2}^2}}{\sin\alpha_M}. \tag{1.8}$$

From the (1.8), it becomes clear that the radial mean squared error value of an aircraft's position determination depends on the variance of LOP measurement's errors and from the α_M angle of LOP crossing. Maximum accuracy of position determination at fixed σ_{l_1} и σ_{l_2} will exist in such case, when LOP cross at the angle $\alpha_M = 90^\circ$ or close to this parameter.

1.7 Coverage Areas of Essential Radio Navigation Systems Types

1.7.1 Methodology of Coverage Area Construction

The coverage area provided by a stationary, ground-based navigation system is that surface area or space volume in which the signals are adequate to permit the aircraft to determine position to a specified level of accuracy. Coverage areas of different radio navigation systems have different sizes and configurations.

Generic methodology of radio navigation systems coverage area's construction includes following stages:

boundary delimitation, within which by condition of operating range, a stable signal reception from ground stations on board of an aircraft is possible. Such factors as transmission power, directional pattern's diagram form of ground-based beacon, signal suppression by the surface's relief or man-made objects, electromagnetic and jamming environment in the coverage area of the system affect the ability or stability of signal receipt. Antenna systems of most ground-based beacons have the emission sector β in vertical surface from 0° to 40°–60° (Fig. 1.24). That is why there is an area in a funnel shape above the beacon (cone of ambiguity or zone of silence), in which the signal from radio beacon is missing. The diameter of this funnel grows with height increase;
aircraft's position determination curve of constant error construction;
coverage area's common part indention, in compliance with the possibility of stable signal receipt and satisfaction of aircraft position determination's accuracy requirements.

Radio navigation system coverage area's boundaries, on the assumption of given accuracy, can be determined using (1.8) for radial mean square error of position determination, assigning acceptable errors' values.

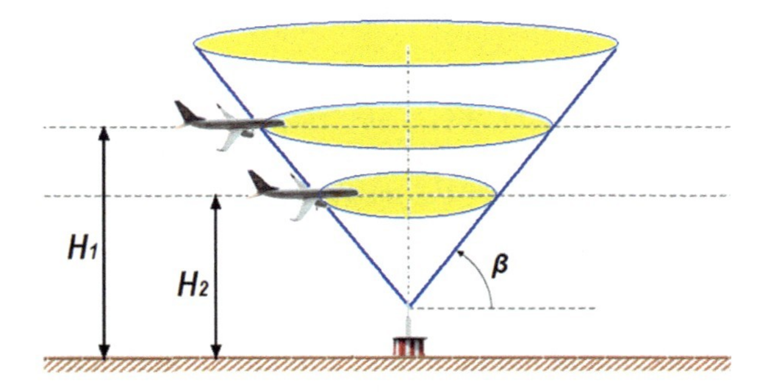

Fig. 1.24 Cone of ambiguity (zone of silence)

1.7.2 Coverage Area of Theta-Theta Navigation System

Let us examine the construction of a theta-theta navigation system's coverage area which consists of two ground-based radio beacons (e.g., radio beacons VOR, situated in *A* and *B* points (Fig. 1.25)).

The following values are outlined in Fig. 1.25: *d*—system's baseline; R_1, R_2—distance from the aircraft to radio beacons *A* and *B*; Θ_1, Θ_2—bearings of aircraft toward radio beacons; α_M—angle of position lines' crossing; *M*—point which matches the true position of an aircraft; *AM*, *BM*—lines of position, determined without errors; *AM′*, *BM′*—lines of position, determined as a result of azimuths' measuring with errors $\delta\Theta_1$ and $\delta\Theta_2$; *M′*—aircraft position, determined as a result of measurements, which is different from the true one by a quantity of radial error *r*; $l_1 = ML$ and $l_2 = MN$—perpendiculars, lined from the *M* point to the lines of position *AM′* and *BM′* accordingly; l_1 and l_2 describe the lines' of position linear errors' measurements in *M* point.

It is supposed that the meridians passing through radio beacons *A* and *B* are parallel.

Taking into consideration that the azimuth measurements' errors $\delta\Theta_1$ and $\delta\Theta_2$ are sufficiently small (for real systems they are not bigger than 2°–3°), linear errors l_1 and l_2 of position lines' determinations satisfy following equations:

$$l_1 = R_1 \sin \delta\Theta_1 \approx R_1 \delta\Theta_1$$
$$l_2 = R_2 \sin \delta\Theta_2 \approx R_2 \delta\Theta_2,$$

where $\delta\Theta_1$ and $\delta\Theta_2$ are measured in [radian].

Then the position lines' error dispersions under such conditions where the expectancies $m_{l1} = m_{l2} = 0$ will be equal to:

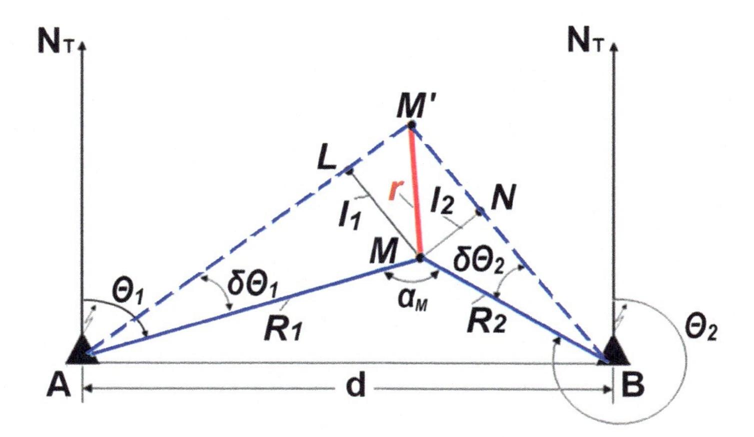

Fig. 1.25 Aircraft's position determination using the theta-theta navigation method

$$\sigma_{l_1}^2 = R_1^2 \sigma_{\Theta_1}^2,$$
$$\sigma_{l_2}^2 = R_2^2 \sigma_{\Theta_2}^2,$$

where $\sigma_{\Theta_1}^2$ and $\sigma_{\Theta_2}^2$ are the azimuth measurements' error dispersions.

Assuming the mean square errors (MSE) of azimuth measurements are equal, e.g., $\sigma_{\Theta_1} = \sigma_{\Theta_2} = \sigma_\Theta$ and mutually independent, we will get the equation for RMS on the basis of (1.8):

$$\sigma_r = \frac{\sqrt{R_1^2\sigma_\Theta^2 + R_2^2\sigma_\Theta^2}}{\sin\alpha_M} = \frac{\sigma_\Theta}{\sin\alpha_M}\sqrt{R_1^2 + R_2^2}. \tag{1.9}$$

From the (1.9), it is seen that the σ_r error depends on the distance to the ground-based stations R_1 and R_2, angle of position lines' crossing α_M and MSE of azimuth measurement σ_Θ.

Let us define $\sigma_{r\min}$ is a minimal RMS value. Supposing $R_1 = R_2 = R$, the value of R distance itself will be expressed through an angle α_M:

$$R = \frac{d}{2\sin\frac{\alpha_M}{2}}.$$

Then the (1.9) will be written as:

$$\sigma_r = \frac{\sigma_\Theta d}{\sqrt{2}\sin\alpha_M \sin\frac{\alpha_M}{2}}. \tag{1.10}$$

To determine the angle α_M, to which the minimal value RMS $\sigma_{r\min}$ is corresponding, we will take the derivate $\mathrm{d}\sigma_r/\mathrm{d}\alpha_M$ and equate it to zero:

$$\frac{\mathrm{d}\sigma_r}{\mathrm{d}\alpha_M} = \frac{\sigma_\Theta d}{\sqrt{2}} \frac{\left(-\cos\alpha_M \sin\frac{\alpha_M}{2} - \frac{1}{2}\sin\alpha_M\cos\frac{\alpha_M}{2}\right)}{\left(\sin\alpha_M \sin\frac{\alpha_M}{2}\right)^2} = 0.$$

The derivate $\mathrm{d}\sigma_r/\mathrm{d}\alpha_M$ is equal to zero if:

$$2\cos\alpha_M \sin\frac{\alpha_M}{2} = -\sin\alpha_M\cos\frac{\alpha_M}{2},$$

or

$$tg\,\alpha_M = -2tg\frac{\alpha_M}{2}.$$

Solving this equation, we will get that $\alpha_M \approx 109^\circ$.

Thereat, the minimal RMS value is defined as:

$$\sigma_{r\min} = 0.01605\sigma_\Theta d,$$

where the σ_{Θ} value is given in [deg].

For instance, at $d = 100$ km and $\sigma_{\Theta} = 1°$ we will get that $\sigma_{r\min} = 1605$ m. In all other points of coverage area, the position's determination error rate will be higher under the same conditions.

To the minimal aircraft's position determination error value the point corresponds that is located on the perpendicular passing through the middle of a baseline.

To determine the boundaries of a coverage area, the curve of constant error must be built in any point where the condition $\sigma_r = \sigma_{radm}$ is executed (σ_{radm}—admissible error). To build a coverage area, the (1.10) might be used; however it is associated with significant calculation difficulties.

In practice, there is an easier method that is used. Let us express the error σ_r through angles φ_1 and φ_2 that describe the position lines' inclination toward the system's baseline (Fig. 1.26a).

Since $\alpha_M = 180° - (\varphi_1 + \varphi_2)$ and:

$$R_1 = d\frac{\sin\varphi_2}{\sin(\varphi_1 + \varphi_2)},$$
$$R_2 = d\frac{\sin\varphi_1}{\sin(\varphi_1 + \varphi_2)},$$

then:

$$\sigma_r = K(\varphi_1, \varphi_2)d\sigma_{\Theta},$$

where:

$$K(\varphi_1, \varphi_2) = 0.017\frac{\sqrt{\sin^2\varphi_1 + \sin^2\varphi_2}}{\sin^2(\varphi_1 + \varphi_2)}.$$

To construct a coverage area, auxiliary tables for $K(\varphi_1, \varphi_2)$ coefficient that depend on the φ_1 and φ_2 values are used. Being given the admissible error σ_{radm} and MSE of azimuth measurement σ_{Θ} the coefficient's value is found for the concrete system with the d baseline as:

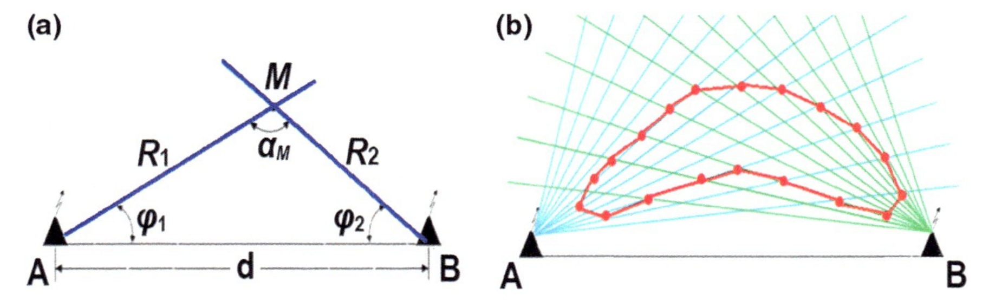

Fig. 1.26 Explaining the method of constructing the coverage area

$$K(\varphi_1, \varphi_2) = \frac{\sigma_{r\,\mathrm{adm}}}{d\sigma_\Theta}.$$

Due to calculated coefficient $K(\varphi_1, \varphi_2)$ for various angles φ_1 two values of another angle φ_2 and φ_2' are chosen from the table. By setting aside lines of position pair-wise at angles φ_1 and φ_2, φ_1 and φ_2', intersection points that belong to the curve of equal accuracy (Fig. 1.26b) are found. Connecting received points of plane curve, we get the boundary of coverage area.

This task can be more effectively solved with the help of computer and appropriate software.

The coverage area of theta-theta navigation system is shown in Fig. 1.27, that is constructed for several $K(\varphi_1, \varphi_2)$ values (at fixed baseline d), considering the directional properties of antenna in vertical surface. For $K(\varphi_1, \varphi_2) = 0.02$ coefficient's value, the coverage area on the image is painted over.

At $K \leq 0.017$, the boundary of a coverage area does not pass the ground stations, taking a form of closed curve. Maximal accuracy that is equal to $\sigma_{r\min}$ value can be attained in one point—O point that is situated on the perpendicular lined through the middle of the baseline.

Outer boundary of the area is normally determined by the set value of position's determination admissible error $\sigma_{r\mathrm{adm}}$, potential accuracy of azimuth measurement σ_Θ, and the operating range of ground radio station, while the inner boundary is determined by the geometrical factor, i.e., accuracy decreasing at the expense of increasing the angle α_M. Admissible error $\sigma_{r\mathrm{adm}}$ increase or azimuth's measurement MSE σ_Θ decrease at fixed baseline of a system d leads to the space (size) extension of a coverage area.

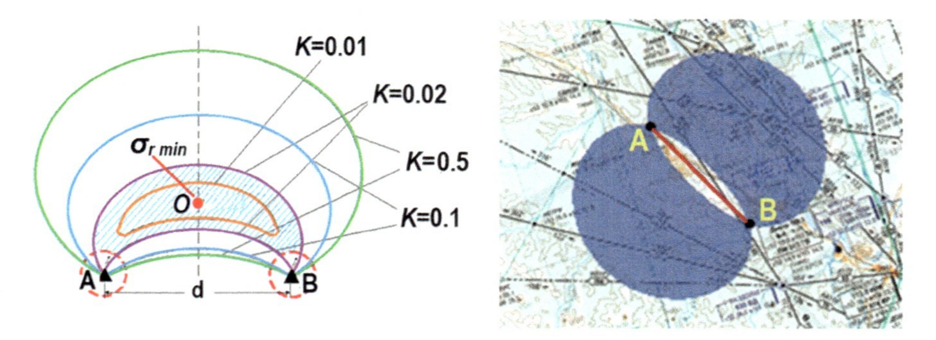

Fig. 1.27 Coverage area of theta-theta navigation system

1.7.3 Coverage Area of Rho-Rho Radio Navigation System

Let us observe the methodology of constructing the coverage area of ranging (rho-rho) radio navigation system that includes the airborne interrogator and two ground-based transponders (e.g., DME-DME) that are located in A and B points with known coordinates (Fig. 1.28). The distance d between ground-based stations forms the baseline of a system, and the airborne interrogator calculates the distance to R_1 and R_2 stations accordingly. Angle ψ_B between two straight lines R_1 and R_2 is called the ***baseline angle*** and is connected with the LOP crossing angle α_M by correlation $\psi_B = 180° - \alpha_M$. Thereat, LOP crossing angle is the angle of tangent lines crossing that are traced to LOP in M point.

Aircraft's position is determined as a crossing point of LOP1 and LOP2 that are defined as R_1 and R_2 radius circles.

Supposing that the distance between the aircraft and the ground-based station is defined by calculating the delay time $\tau = t_{rec} - t_{int}$ between the moment of interrogation signal's emission t_{int} and the moment of reply signal's reception t_{rec} by the interrogator. Then the following equation can be written:

$$R = \frac{c}{2}\tau,$$

where c is the speed of radio-wave propagation.

At constant and known speed of radio-wave propagation, the error of distance determination is:

$$\delta R = \frac{c}{2}\delta\tau,$$

where $\delta\tau$ is the error of delay time calculation using the interrogator facilities.

The error of distance measurement in range system is the error of LOP measurement itself. Supposing that the mathematical expectation of error $M\{\delta\tau\} = 0$ and the error of delay time measurement is under the control of Gaussian law, we get the following equation:

$$\sigma_R = \frac{c}{2}\sigma_\tau,$$

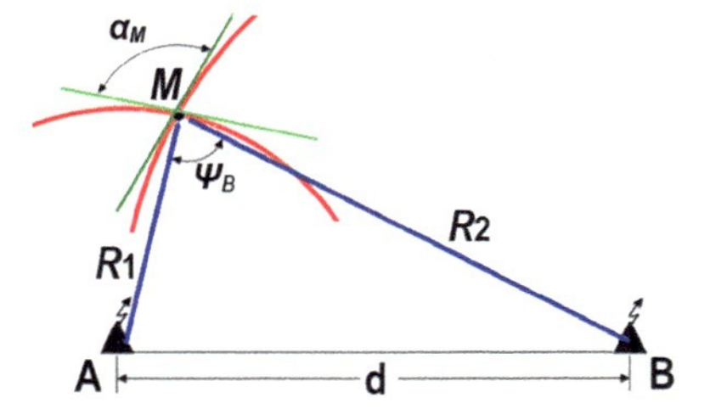

Fig. 1.28 Aircraft's position determination using the rho-rho method

where σ_R, σ_τ are the MSE of distance and delay time measurements.

Then the LOP1 and LOP2 determination accuracy is characterized by following equations:

$$\sigma_{l_1} = \sigma_{R_1} = \frac{c}{2}\sigma_{\tau_1},$$
$$\sigma_{l_2} = \sigma_{R_2} = \frac{c}{2}\sigma_{\tau_2}.$$

Normally, the measurements of time delays are of equal accuracy, i.e., $\sigma_{\tau_1} = \sigma_{\tau_2} = \sigma_\tau$ and in accordance $\sigma_{R_1} = \sigma_{R_2} = \sigma_R$. Then, taking into consideration (1.8), RMS of aircraft's position determination by rho-rho system will be as:

$$\sigma_r = \frac{\sigma_R\sqrt{2}}{\sin\alpha_M} = 0.7\frac{c\sigma_\tau}{\sin\alpha_M}. \tag{1.11}$$

The task of coverage area border finding is reduced to the construction of constant accuracy curve σ_r = const, which satisfies the condition $\sigma_r \leq \sigma_{radm}$, where σ_{radm} is the value of acceptable RMS of aircraft's position determination within the coverage area of system that is being defined by requirements to the accuracy of aircraft's navigation supplement.

Proceeding from the (1.11), σ_r = const at:

$$\sin\alpha_M = \frac{\sigma_r}{\sigma_{radm}}\sqrt{2} = \text{const},$$

i.e., in every point of constant accuracy curve the angle of LOP crossing α_M is the constant value.

Graphical constructions using (1.11) show that the curves of constant accuracy in rho-rho system are the circles that are based upon the d baseline as if upon the chord, with the central angle $2\alpha_M$. Considering that the radius r of these circles can be found for various σ_{radm} values from the geometrical constructions (Fig. 1.29).

It is seen in Fig. 1.29 that the circle radius that corresponds to the set value σ_{radm} is defined by the following equation:

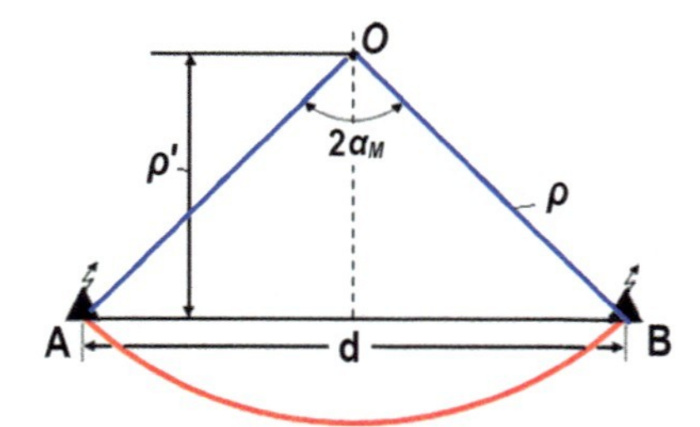

Fig. 1.29 Circle radius finding

$$\rho = \frac{d}{2\sin\alpha_M}.$$

The center of circle (O point) is located on the perpendicular that is traced through the center of a baseline at the distance $\rho' = \rho\cos\alpha_M = \frac{d}{2}\mathrm{ctg}\,\alpha_M$ from the baseline.

From the (1.11), it implies that the minimum RMS $\sigma_{r\min}$ value is achieved at $\alpha_M = 90°$, i.e., on the circle for which the d baseline is the diameter and makes $\sigma_{r\min} = \sqrt{2}\sigma_R$. At the disposal from the baseline, the angle α_M is decreasing which leads to the position determination's accuracy degradation. At approaching the baseline, the α_M angle becomes bigger than 90° which also lead to the degradation in aircraft's position determination's accuracy.

Let us construct the coverage area of rho-rho radio navigation system for the following case $\sigma_{r\mathrm{adm}} = 2\sigma_{r\min} = 2\sqrt{2}\sigma_R$ (Fig. 1.30). According to the (1.11), such value $\sigma_{r\mathrm{adm}}$ corresponds to the value $\sin\alpha_M = 1/2$ and the angles' values $\alpha_{M1} = 30°$ and $\alpha_{M2} = 150°$.

The coverage area of rho-rho radio navigation system at omnidirectional antenna emission of the ground-based radio beacons is symmetrical toward the baseline, that is why in Fig. 1.30 only the half of coverage area is shown.

Aircraft's position determination is impossible at its being on the baseline and over baselines in rho-rho system, since the LOP is parallel in such cases.

The square (size) of coverage area increases at MSE σ_R of range measurement decrease and at the increase of the set value of acceptable position determination RMS $\sigma_{r\mathrm{adm}}$.

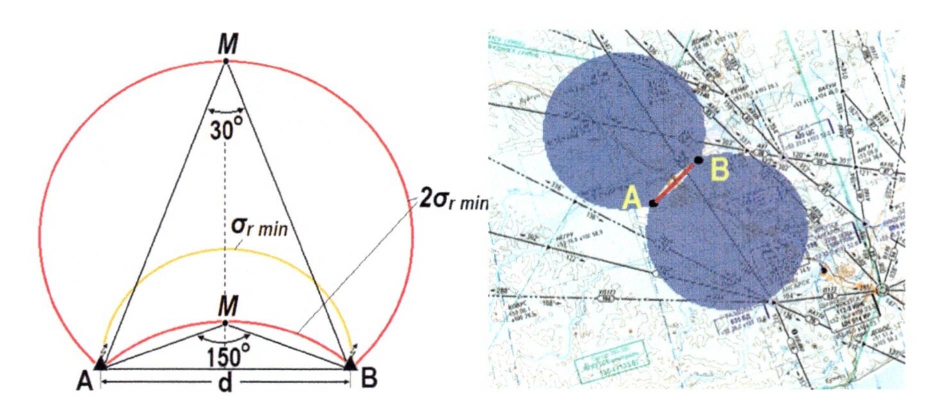

Fig. 1.30 Coverage area of rho-rho radio navigation system

1.7.4 Coverage Area of Rho-Theta Navigation System

Let us examine the rho-theta navigation system, the top part of which is formed by azimuthal and ranging radio beacons (RB) situated in the same point (e.g., VOR-DME). The position of an aircraft is determined as a crossing point of two LOPs: lines of uniform azimuth LOP1 and lines of equal range LOP2 (Fig. 1.31) that are defined using azimuthal and ranging channels.

At the rho-theta navigation system radio beacons' allocation in one point, the angle of LOP crossing is $\alpha_M = 90°$.

At mutually independent azimuth and range measurements' errors, that normally takes place in practice, RMS of position determination fills the equation:

$$\sigma_r = \sqrt{R^2\sigma_\theta^2 + \sigma_R^2}, \tag{1.12}$$

where $R\sigma_\Theta$—MSE of LOP1, σ_R—MSE of LOP2 determination.

Equation (1.12) can be used at building the rho-theta navigation system's coverage area on the surface, wherefore it is required to set the acceptable RMS value of an σ_{radm} error of aircraft's position determination. Thereat, in every point of the constant accuracy curve the following condition must be fulfilled:

$$\sigma_{radm} = \sqrt{R^2\sigma_\Theta^2 + \sigma_R^2} = \text{const}.$$

In this equation, range R is the constant, the choice of which provides the condition's (1.12) fulfillment. Therefore, the curve of constant accuracy will be defined as the R radius circle, which center is situated in the point of radio beacons location.

Circle radius which can be described as the curve of constant accuracy (i.e., border of coverage area) can be found from the (1.12) as:

$$R = \frac{1}{\sigma_\Theta}\sqrt{\sigma_{radm}^2 - \sigma_R^2}. \tag{1.13}$$

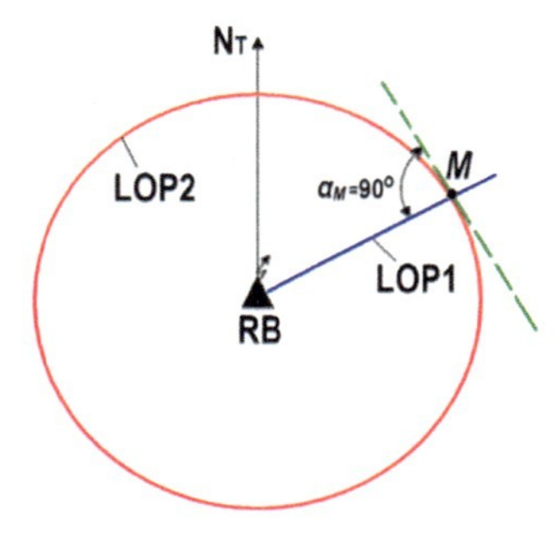

Fig. 1.31 Position determination using the rho-theta system

Maximum operating distance of azimuth and range measurement channels of VOR-DME-type rho-theta navigation system is defined principally by the conditions of wave propagation. Since the system functions in VHF band, maximum operating distance is defined by the distance of the straight visibility as:

$$D_{\max}[\mathrm{km}] = 4.12\left(\sqrt{h_1 + h_2}\right),$$

where h_1 [m]—height of receiving antenna that is equal to the height of an aircraft's flight, h_2 [m]—transmitting antenna's height.

In stiff terrain that has obstructions for rectilinear radio-wave propagation, the definition of $D_{\max}$ is conducted by sectors, within which the conditions of radio-wave propagation can be considered equal, taking into account the aircraft's flight heights, obstructions, and radio beacon's antenna, as well as its mutual position.

As a result of coverage area's R radius calculations according to (1.13) and considering the maximum operating distance $D_{\max}$ that is limited to the borders of straight visibility, the coverage area's radius R of rho-theta navigation system is defined. It is equal to the minimal from the R or $D_{\max}$ values (Fig. 1.32).

Once this is done, the circle centered in point of radio beacon's location (so-called "silence zone" that is caused by limited field of radio beacons' emission in the vertical surface) is excluded from the area which is narrowed by R radius. Radius of this circle is normally equal to the height of aircraft's flight. Remaining area describes the coverage area of rho-theta navigation system (painted area in Fig. 1.32).

Since the outer radius of system's coverage area is defined by a value that is calculated using the (1.13) and the distance of straight visibility (the smaller of these two values), the increase of set acceptable RMS error of position determination $\sigma_{r\mathrm{adm}}$ and decrease of azimuth measurement MSE σ_{Θ} and range σ_R does not always lead to the increase of coverage area's size. The coverage area's size increase does not always result from the flight height's increase as well.

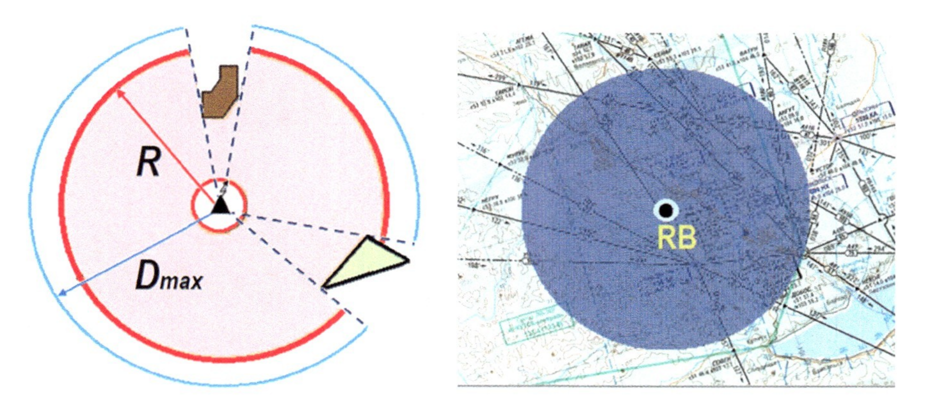

Fig. 1.32 Coverage area of rho-theta navigation system

1.7.5 Coverage Area of Hyperbolic Fixing (Range Difference) Navigation System

In hyperbolic fixing navigation system, the aircraft's position is determined as a point of two lines of position, LOP1 and LOP2 (hyperboles) crossing, that are received for two pairs of ground-based stations, e.g., *A* and *B*, *A* and *C*, which coordinates are known (Fig. 1.33).

In Fig. 1.33 ψ_{B1}, ψ_{B2} are the baseline angles; α_M—angle of LOP crossing (it is defined as an angle between the tangents *PQ* and *RS*, traced to the hyperboles in the *M* point).

LOP is defined by the range difference $R_B - R_A$, $R_C - R_A$ till the ground-based stations by measuring the difference in moments' receipt of impulse signals. Thereat, the timeout of signal receipt moments for the pair of *A* and *B* stations measured on board is defined by the following equation:

$$\tau_{AB} = \frac{R_B - R_A}{c}.$$

Therefrom the range difference is $R_{AB} = |R_B - R_A| = c\tau_{AB}$.

The error of $\delta\tau_{AB}$ delay measurement leads to a range difference measurement error $\delta R_{AB} = c\delta\tau_{AB}$, which in its turn leads to the LOP displacement against the true one at a distance of:

$$l = \frac{\delta R_{AB}}{2\sin\frac{\psi_{B1}}{2}}.$$

Assuming the error expectation is $M\{l\} = 0$, we get the equation for MSE LOP determination:

$$\sigma_{l_{AB}} = \frac{\sigma_{R_{AB}}}{2\sin\psi_{B1}} = \frac{c\sigma_{\tau_{AB}}}{2\sin\psi_{B1}}. \tag{1.14}$$

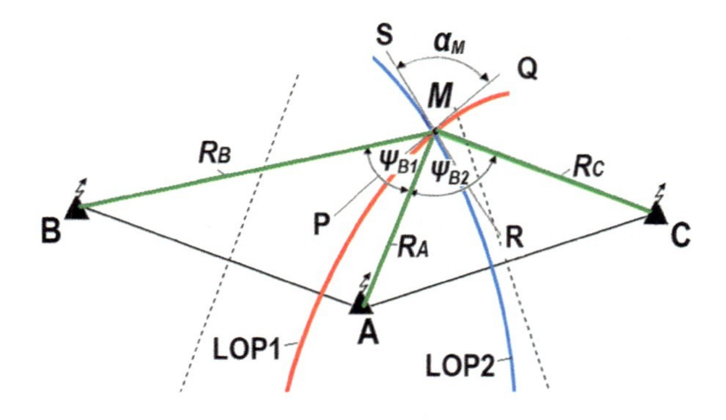

Fig. 1.33 Position determination using hyperbolic fixing methodology

LOP determination error, which is caused by A and C stations, has similar pattern. As a result, in hyperbolic fixing navigation system the error of LOP determination depends on the error of range difference measurement (or the temporary interval between moments of ground-based stations' impulse reception) and the baseline angle.

By applying the (1.14) to the expression (1.8) for RMS error, we get the following:

$$\sigma_r = \frac{c\sqrt{\sigma_{\tau_{AB}}^2 \sin^2 \frac{\psi_{B2}}{2} + \sigma_{\tau_{AC}} \sin^2 \frac{\psi_{B1}}{2}}}{2 \sin \alpha_M \sin \frac{\psi_{B1}}{2} \sin \frac{\psi_{B2}}{2}}. \quad (1.15)$$

Equation (1.15) analysis shows that in hyperbolic fixing navigation system, the accuracy of position determination gets worse in the following situations:

at the distance increase from the station baseline which is caused by the reduce of baseline angles;
at being ahead of the baselines when one of the baseline angles is close or equal to zero;
at being at the points of coverage area where the LOP's crossing angle is pretty small.

The configuration and the size of coverage area depend on the angle value between the baselines. Optimal angle value, at which the surface of coverage area is close to the maximum, is $\beta = 90°–120°$.

For the hyperbolic fixing radio navigation systems coverage area building according to the (1.15) computer programs are used. Thereat, it is usually considered that $\sigma_{\tau AB} = \sigma_{\tau AC} = \sigma_{\tau}$.

Then, the equation for RMS error can be written in the following form:

$$\sigma_r = c\,\sigma_\tau\, K\{\psi_{B1}, \psi_{B2}, \alpha_M\},$$

where

$$K(\psi_{B1}, \psi_{B2}, \alpha_M) = \frac{\sqrt{\sin^2 \frac{\psi_{B1}}{2} + \sin^2 \frac{\psi_{B2}}{2}}}{2 \sin \alpha_M \sin \frac{\psi_{B1}}{2} \sin \frac{\psi_{B2}}{2}}$$

is an index which is called the system's geometric dilution of precision which takes into account the impact of mutual aircrafts' positions and ground-based stations on the accuracy of position determination.

For the coverage area's construction, the RMS error acceptable value is set $\sigma_{r\text{adm}}$ and by computer calculations the curve is plotted:

$$K(\psi_{B1}, \psi_{B2}, \alpha_M) = \frac{\sigma_{r\text{adm}}}{c\sigma_\tau} = \text{const.}$$

Coverage areas in hyperbolic fixing navigation system for different amount of ground-based stations are illustrated in Fig. 1.34.

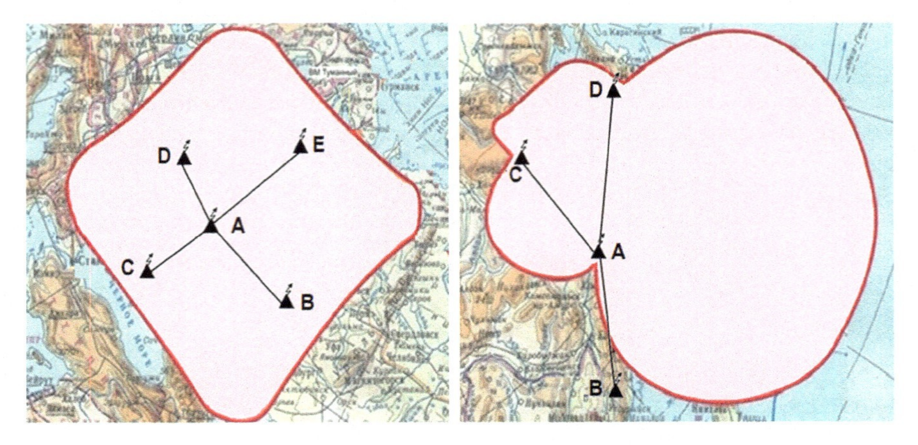

Fig. 1.34 Coverage area in hyperbolic fixing navigation system for different amount of ground-based stations

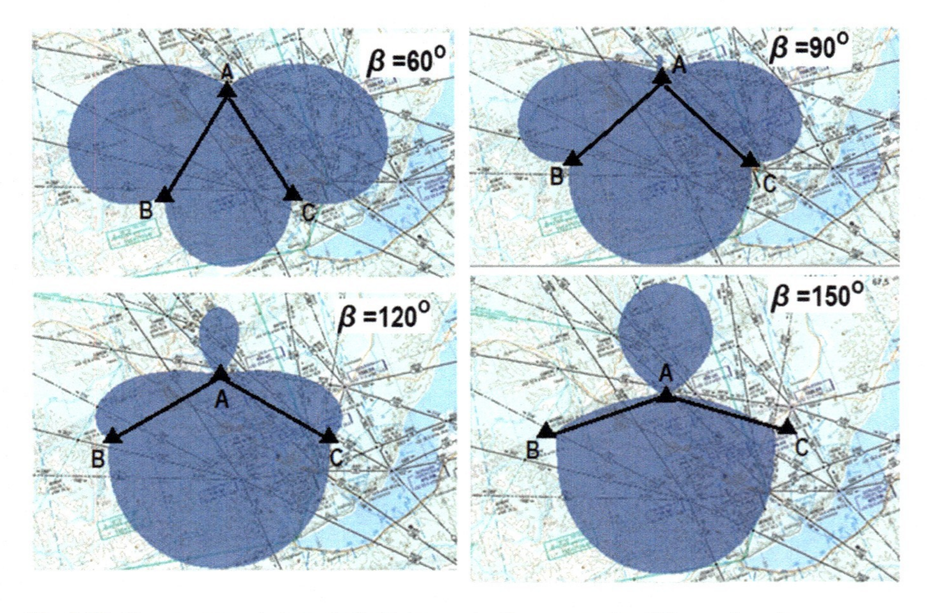

Fig. 1.35 Coverage area in hyperbolic fixing navigation system for different β angles

Coverage areas in hyperbolic fixing navigation system for different β angle values are illustrated in Fig. 1.35.

Coverage area surface grows at the RMS error of position determination set value $\sigma_{r\text{adm}}$ increase and at the increase of range difference measurement accuracy (at the MSE σ_{τ} decrease).

1.8 Aeronautical Consumers' Requirements to the Navigation Accuracy

Air traffic's continuous growth of intensity and density leads to the necessity of the airspace capacity increase. For this, the task of its optimal usage must be solved on basis of new air traffic organization, technologies, and technical navigation aids concepts' implementations.

Resolution of this and other tasks which are oriented on the aviation's transport efficiency increase becomes possible at the conception of performance-based navigation (PBN) implementation.

"The PBN concept is a global initiative by the International Civil Aviation Organization (ICAO). At the 37th ICAO General Assembly, a resolution was adopted urging implementation of the PBN concept. The Assembly urges all states to implement RNAV and RNP air traffic services (ATS) routes and approach procedures in accordance with the ICAO PBN concept laid down in the performance-based navigation (PBN) manual (Doc 9613)" [2].

In the heart of PBN conception, which is based on the area navigation principles area navigation (RNAV) and required navigation performance (RNP), lies the transition from sensor-based navigation to the navigation which is based on characteristics (accuracy first of all). At the same time, by the accuracy we understand not only the estimation of position accuracy of an aircraft, characterized by Navigation System Error (NSE), but the accuracy of its retention in a certain area, limited by longitudinal and lateral deviation from a given flight path. This area is defined by the Total System Error (TSE) which is usually computed as a root sum square of the NSE, the flight technical error (FTE) and the position definition error (PDE).

The required accuracy is achieved by usage of such radio navigation systems as VOR/DME, DME/DME, ILS, GNSS—as sources of navigation information, and flight management system (FMS)—as device of its procession together with the data of the dead reckoning systems.

"RNAV and RNP systems are fundamentally similar. The key difference between them is the requirement for onboard performance monitoring and alerting. A navigation specification that includes a requirement for onboard navigation performance monitoring and alerting is referred to as an RNP specification. RNAV specifications do not require onboard performance monitoring and alerting. An area navigation system capable of achieving the alerting performance requirement of an RNP specification is referred to as an RNP system.

The PBN concept specifies that aircraft area navigation system performance requirements be defined in terms of the accuracy, integrity, continuity, and functionality, which are needed for the proposed operations in the context of a particular airspace concept; availability is a function of the navigation signal in space. This represents a shift from sensor-based to performance-based navigation. Performance requirements are identified in navigation specifications, which also identify the choice of navigation sensors and equipment that may be used to meet the performance requirements" [2].

There are two kinds of navigation specification: RNAV and RNP.

RNAV is defined as a method of navigation which permits aircraft operations on any desired track within the coverage area of radio beacons (ground or satellite based), or within the limits of accuracy of a dead reckoning navigation system.

RNAV system is any system that allows the aircraft to be navigated to the required level of accuracy without the requirement to fly directly over ground-based radio beacons. In other words, RNAV methods and systems' usage allow to transfer from en route flights that are set through ground-based navigation beacons (FROM or TO the beacons) to the en route flights that are set in the navigation base FMS by coordinates of WP in the global coordinate system.

RNAV specifications have been developed to support existing capabilities in aircraft equipped with area navigation systems.

RNP is a type of performance-based navigation that allows an aircraft to fly a specific path between two 3D-defined points in space.

Navigation specifications are set with the prefix value RNAV *X* (e.g., RNAV 2) and RNP *X* (e.g., RNP 2).

"*X*" points at the accuracy of lateral navigation in nautical miles which should be held during at least 95% of flight time by all aircrafts that conduct flights within the set airspace, en route, or according to the flight pattern.

Published navigation specifications are [3]:

RNAV 10, RNAV 5, RNAV 2, RNAV 1;
RNP 10, RNP 4, RNP 2, RNP 1, Advanced RNP, RNP APCH, RNP AR APCH, RNP 0.3.

It should be noted that only the RNP specifications are used for the approach and landing phase.

1.9 Summary

Aeronautical radio navigation, its methods, and technical aids are fast-growing areas of science and technics. Constant air traffic intensity and density growth increase the role of aeronautical radio navigation in providing the regularity of air traffic and flights safety. Air navigation aids are used on all stages of flight—from aircraft's takeoff till its landing on the planned airfield, and in the recent years—even during its taxing on the airfield's surface.

Overview on technical aids of ground base navigation is given in this book, which operate together with aircraft's onboard aids and are developed for solving the navigation tasks in the airfields' area as well as on the air routes. Many of the systems which are described in the book have been successfully used in the civil aviation for a lot of years. These aids can be considered traditional. Since the traditional aids have proven their efficiency, in the mid-term perspective its use in the civil aviation will be continued. Herewith, the process of improving its technical and performance characteristics will also be continued.

However, the potential of some traditional navigation systems in terms of accuracy and functionality is depleted or is close to it. That is why the attention of aviation community and science is focused on the development and use of new technologies and technical navigation aids which are based on the satellite navigation systems.

In the first chapter of the book together with traditional, new directions of technical aid's development and methods of navigation are overviewed. Chapters 2–4 are focused on traditional means of navigation (systems NDB, VDF, VOR, DME, ILS), and Chaps. 5 and 6—on relatively new and perspective technologies and technical aids (GNSS, GLSS, MLAT).

1.10 Further Reading

Requirements for aviation consumers to the accuracy of positioning on different aircraft's flight stages, to the characteristics of concrete types of technical aids of radio navigation, can be found in normative documents of ICAO [2–6] as well as in [1, 7]. In [1, 7], one can find the definition (explanation) of main navigation aids' characteristics, and get an insight on radio navigation aids used in the USA and Russia and its development perspectives.

Common questions on the theory of air and flight navigation, navigation parameters' characteristics are described in such books as [8, 9].

More detailed information on coordinate systems used in navigation, detailed analysis of their characteristics, and coordinate transformations are in [10–14].

More detailed information about perspectives of air navigation development by using the block modernization can be taken from [15].

References

1. 2017 Federal Radionavigation Plan. United States. Dept. of Defense; United States. Dept. of Homeland Security; United States. Dept. of Transportation; National Technical Information Service. Download from https://rosap.ntl.bts.gov/view/dot/32801
2. Performance-based navigation (PBN) Operations Plan June 2017. Download from www.navcanada.ca/EN/products-and-services/Documents/Operations%20Plan_Performance%20Based%20Navigation_E_WEB.pdf
3. Performance based navigation (PBN) manual. Doc 9613 AN/937. ICAO. Montreal, 4th edn (2013). Download from https://docslide.net/documents/icao-pbn-manual-fourth-edition-pbn-manual-fourth-edition-icao-doc-9613-final.html
4. Introducing Performance Based Navigation (PBN) and Advanced RNP (A-RNP). European Organization for the Safety of Air Navigation (EUROCONTROL). January, (2013). Download from https://www.eurocontrol.int/sites/default/files/publication/files/2013-introducing-pbn-a-rnp.pdf
5. Annex 10 to the Convention on International Civil Aviation. Aeronautical Telecommunications, vol I. Radio Navigation Aids, 6th edn (2006)

6. Manual on Required Navigation Performance (RNP) (Doc 9613-AN/937) (1999) International civil aviation organization, 2nd edn. Download from http://www.wing.com.ua/images/stories/library/ovd/9613.pdf
7. Radionavigacionnyj plan Rossijskoj Federacii. [Radionavigational plan or Russia Rederation]. Utverzhden prikazom Minpromtorga Rossii ot 28.06. 2015. №2123 (in Russian). Download from http://docs.cntd.ru/document/456034328
8. Skrypnik ON (2014) Radionavigacionnye sistemy vozdushnyh sudov [Radionavigation systems of aircrafts]. Moscow. INFRA-M (in Russian)
9. Saraiskii YN, Aleshkov II (2013) Aeronavigaciya. Chast 1. Osnovy navigacii i primenenie geotehnicheskih sredstv. [Airnavigation. Part 1. Fundamentals of radio navigation and the use of geotechnical means], 2nd edn. Sankt-Petersburg. SPbUGA (in Russian)
10. Doc 9674 AN/946 (2002) World Geodetic System-1984 (WGS-84) Manual. International Civil Aviation Organization, 2nd edn. Download from http://gis.icao.int/eganp/webpdf/REF08-Doc9674.pdf
11. Moritz H (1980) Geodetic reference system 1980. Download from https://ru.scribd.com/document/277066804/Geodetic-Reference-System-1980
12. ITRF (2008) Download from http://itrf.ensg.ign.fr/ITRF_solutions/2008
13. Parametry Zemli 1990 goda (PZ-90.11). [The Earth's parameters (PZ-90.11)]. Spravochnyj dokument. Voenno-topograficheskoe upravlenie General'nogo shtaba Vooruzhennyh sil Rossijskoj Federacii. Moscow (2014) (in Russian)
14. Postanovlenie Pravitel'stva Rossijskoj Federacii ot 28.12.2012 №1463. O edinyh gosudarstvennyh sistemah koordinat. Download from http://gis-lab.info/docs/law/statecoord-2012.doc (in Russian)
15. Global Air Navigation Plan 2016–2030. Doc 9750-AN/963, 5th edn (2016). Download from https://www.icao.int/publications/Documents/9750_5ed_en.pdf

Chapter 2
Direction-Measuring Short-Range Navigation Systems

Direction-measuring radio navigation systems are made for setting or determining the directions in space. This allows to perform aircrafts' positioning on the set points of route (airfield, WP) and control their position in the air space.

Direction-measuring system contains of non-directional radio beacon (NDB) and radio-bearing—radio direction finder (RDF). Radio beacon's antenna system can have directional or non-directional emission. Polar diagram of RDF antenna system's always has directional qualities.

Direction-measuring systems of short-range navigation include guidance ("homing") systems of medium-frequency band consisting of NDB and automatic direction finder (ADF) and radio-bearing systems of VHF band.

NDB and direction-finding systems became the first radio navigation aids used in the aviation. This happened in the beginning of the twentieth century. Nowadays, direction-finding systems play double role for other, more accurate radio navigation systems. However, on the remote continental air routes and at landing on the airfields which are situated in underpopulated and underdeveloped regions (e.g., Siberia and Russia Far East, high latitudes), they often are the only available radio navigation aid.

Radio direction finders play important role at providing the observation (surveillance) of aircrafts on air routes and in airfields' areas.

Section 2.1 gives common characteristics of NDB and direction-finding systems.

Section 2.2 gives an overview on NDB: their purpose and tasks solving on aircraft's navigation and air traffic control, special aspects of route and airfields' radio beacons are shown, special aspects of their placement in the airfield's area by regulations accepted in Russia and regulations, recommended by ICAO. This section shows simplified block diagram of a radio beacon and analyzes its work in the main modes of signal emission. As an example of practical realization, Russian-made radio beacon RMP-200 is described: its construction, main characteristics, block and functional diagrams, and operating principle of the radio beacon.

Section 2.3 describes automated radio direction finders: their purpose, main characteristics, and their placement in the airfield's area. Simplified block diagrams and operating principle of amplitude, amplitude-phase, Doppler, and differential Doppler

O. N. Skrypnik, *Radio Navigation Systems for Airports and Airways*,
Springer Aerospace Technology, https://doi.org/10.1007/978-981-13-7201-8_2

radio direction finders are shown. Main mathematical equations that describe processes of navigation information processing are shown. As an example of practical realization, Russian-made radio direction finder DF-2000 is described: its construction, main characteristics, functional diagram, and operating principle.

2.1 General Characteristics of Short-Range Navigation Systems

The direction-measuring short-range navigation systems are designed to solve aircraft navigation problems by setting or determining directions in space. This enables to guide an aircraft in the target waypoints (airdrome, turning, and way points), to control the aircraft attitude in the airspace from ATC towers.

A direction-measuring navigation system comprises a transmitting (radio beacon) and a receiving (radio direction finder) devices. The radio beacon can have an antenna system with directional or non-directional radiation. The antenna system of a direction finder (DF) always possesses directional properties.

The direction-measuring short-range navigation systems include:

direction-finding ("homer") systems of the low-frequency (LF) and medium-frequency (MF) bands;
direction-finding systems of the very high frequency (VHF) band.

ICAO has assigned a frequency band of 190–1,750 kHz for "homer" systems to operate, but some countries use mainly the lower part of the band. In Russia, a frequency band of 150–1,750 kHz is assigned for "homer" systems. Ground-based transmitters (non-directional beacons, NDBs) and onboard direction finders (automatic direction finders, ADFs) are included in "homer" systems.

Direction-finding systems of the VHF band operate in a frequency band of 118–137 MHz assigned for onboard VHF radio stations or 220–400 MHz (in particular cases). The DF system comprises aircraft VHF radio stations and ground-based direction finders (DF, VDF).

2.2 Non-directional Beacons

2.2.1 Application and Missions

The NDB is a ground-based transmitter which transmits vertically polarized radio signals in all directions (hence the name) and is designed to determine directions to it in space. The vertically polarized signal is needed to create a desired antenna pattern of the ADF antenna system. A loop antenna has such antenna pattern with two defined directions of zero reception (Fig. 2.1).

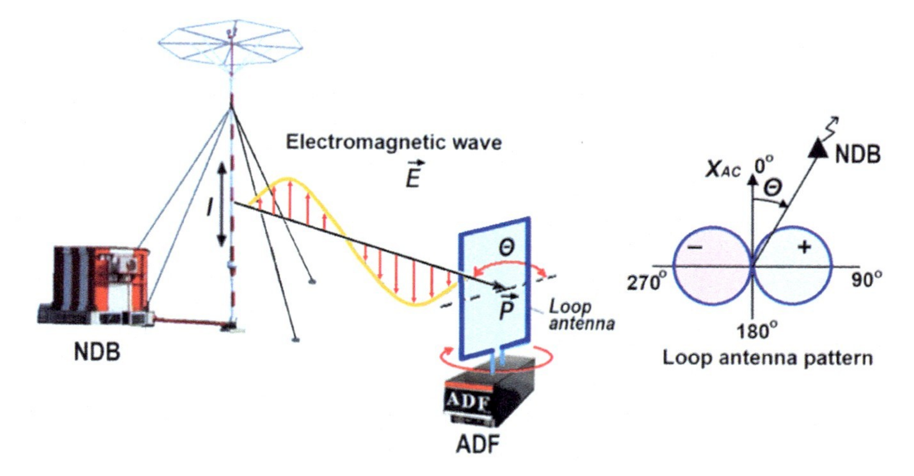

Fig. 2.1 Concerning the NDB-ADF system operation

If a loop aerial is placed in the electromagnetic field of the transmitted radio wave, a voltage will be generated in the vertical elements of the loop. Output voltage of loop aerial defined of the phase difference of the wave in each of the vertical elements. As the loop is turned, the voltage induced will decrease until it becomes zero when the loop plane is perpendicular to the radio-wave direction of propagation.

When an aircraft's ADF receives an NDB's signal, the direction of the NDB will be determined. The ADF can also locate transmitters (broadcast stations) in the standard AM medium-wave broadcast band and, hence, to point in their direction (Fig. 2.2).

ADF determines the direction or bearing to the NDB station relative to the aircraft. Relative bearing Θ is a counted clockwise angle in horizontal plane between the longitudinal axis of the aircraft X_{ac} and the direction to the NDB. Information about the relative bearing may be displayed on a relative bearing indicator (RBI) (Fig. 2.3), on a radio magnetic indicator (RMI) or a navigation display (ND). It also can be sent to the FMS of modern aircraft.

The NDB emits VHF oscillations modulated with an identification signal (radio beacon call sign) or a voice message. The NDB can also emit a continuous unmodulated signal (only a carrier).

As a rule, each NDB is identified by a one-, two-, or three-letter Morse code call sign. However, some countries use an alphanumeric call sign (one letter and one number). Tone amplitude modulation (amplitude-shift keying) is used for audible distinction of signals from different NDBs. In this case, a carrier is modulated with either 400 or 1,020 Hz.

NDB signals propagate as a ground wave and bend around the earth's surface. Their propagation distance can reach several hundred kilometers and depends upon the power of a transmitter, efficiency of an antenna (of its effective height), and

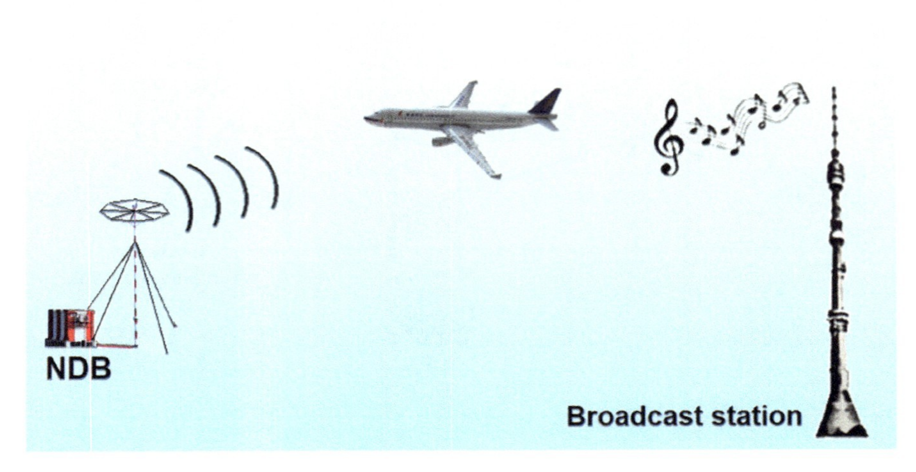

Fig. 2.2 Types of transmitters taken a bearing

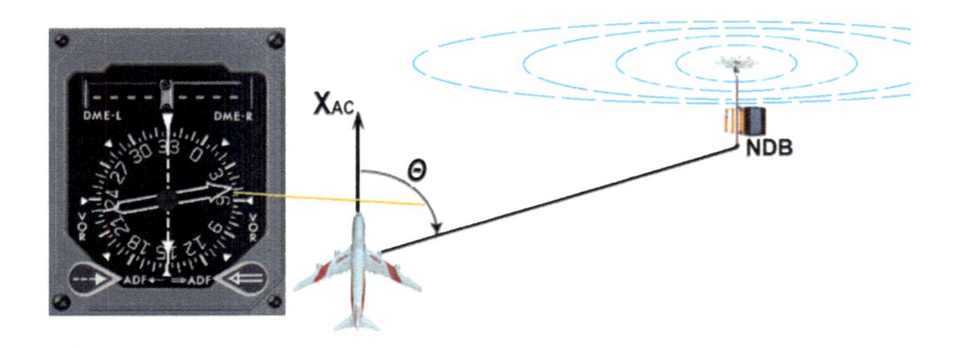

Fig. 2.3 NDB and ADF cooperation

coherence of transmitter output impedance and wave impedance of the antenna. The propagation distance of the signal is also influenced by the condition of a radio path (weather conditions, underlying terrain features, interference environment).

NDB antennas are usually too short for resonance at the frequency they operate—-typically perhaps 20 m length compared to a wavelength around 1,000 m. Therefore, they require a suitable matching network that may consist of an inductor and a capacitor to increase the effective height of the antenna. T-antennas and umbrella-loaded vertical antennas (Fig. 2.4) are mainly used in NDBs. The umbrella-like structure is

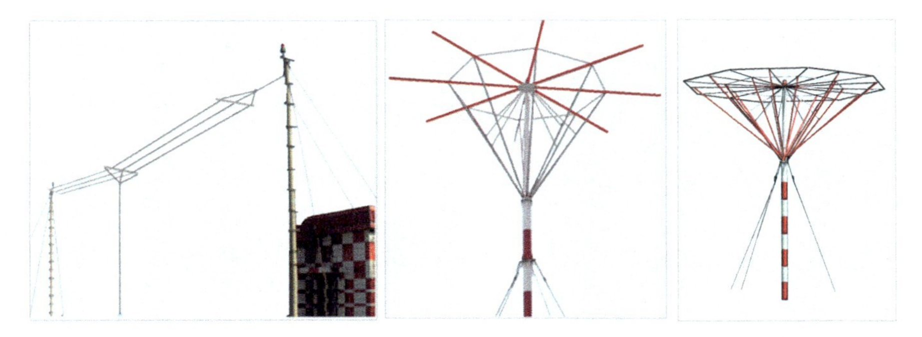

Fig. 2.4 NDB antennas

designed to add loading at the end and improve the radiating efficiency of an antenna. A ground plane or counterpoise is usually connected underneath the antenna.

A "cone of silence" exists overhead the NDB antenna during which the aircraft does not receive any signals. The diameter of the cone increases with aircraft height.

According to their missions and location, the following types of NDBs are distinguished:

en route (separately located) NDBs, used to mark airways and WP. These have a range of 150 km or more. They are used for homing, holding, en route, and airways navigation;
approach (aerodrome) NDBs.

Approach NDBs used in conjunction with an Instrument Landing System (locator beacons). These are low-powered NDBs used for airfield or runway approach procedures and colocated with the outer and middle markers of an ILS (LOM—Locator Outer Marker, LMM—Locator Middle Marker). They normally have ranges of 20–50 km and may only be available during an aerodrome's published hours of operation.

In the medium term, it is expected to stop operating NDBs in the interest of civil aviation. However, nowadays extensive NDB coverage in the USA is provided by 1,260 ground stations, of which FAA operates more than 600. In Russia, NDBs also remain valid in the ATC system; all airdromes and airways are equipped with them.

The NDB-ADF system enables to solve the following problems:

en route navigational bearings with NDB call sign read-through and indication of relative bearings;
air navigation (to fly directly TO the NDB with $\Theta = 0$ and FROM the NDB with $\Theta = 180°$) (Fig. 2.5, indication on the RMI is shown);
determining the compass heading to an NDB station (in a no wind situation), i.e., taking the relative bearing between the aircraft and the radio beacon, and adding the magnetic heading of the aircraft;
NDB approaches (an approach with NDB as the only navigation aid is called a non-precision approach) (Fig. 2.6, indication on the RBI is shown, race-track procedure);

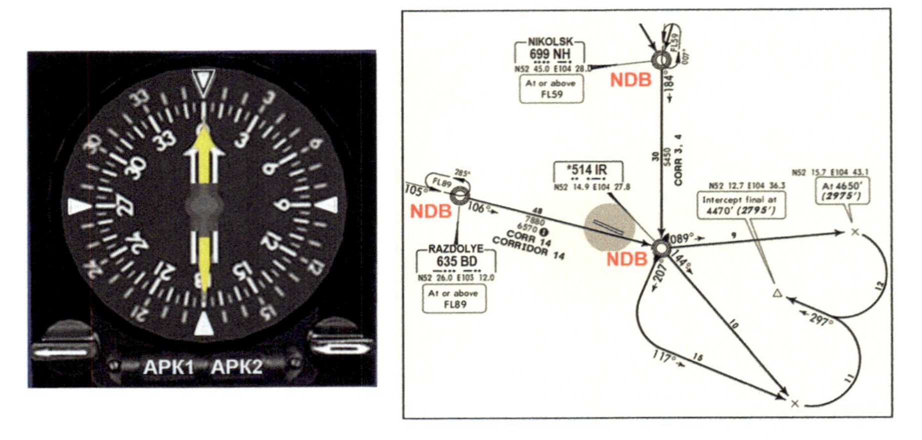

Fig. 2.5 Indication on the RMI and flight pattern “to NDB”

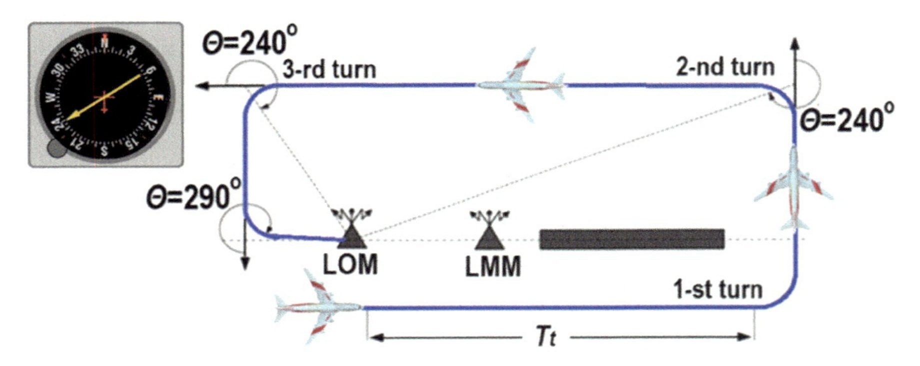

Fig. 2.6 Indication on the RBI and race-track procedure

position fixing an aircraft by bearings taken from two NDBs with known coordinates (Fig. 2.7, indication on a dual-needle RMI is shown). This usage is important in situations where other navigational equipment has failed or has low accuracy, for example, when flying near the North Pole;

other information transmitted by an NDB. So, in an emergency, i.e., air-ground-air communication failure, an air traffic controller using a communication function, may modulate the carrier with voice by microphone. The pilot uses his ADF receiver to hear instructions from the tower.

NDBs are most commonly used as markers or “Compass Locators” for an ILS approach or standard approach. NDBs may designate the starting area for an ILS approach or a path to follow for a standard terminal arrival procedure (STAR).

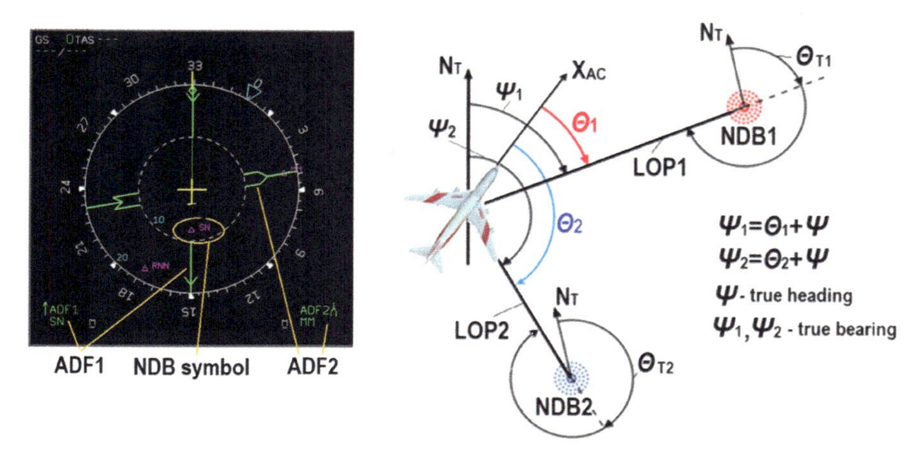

Fig. 2.7 Indication on the ND and the principle of locating by bearings taken from two NDBs

2.2.2 NDB Placement

There are a pretty large number of airdromes having a runway equipped with the landing system in which an NDB is the only navigation aid. As a rule, these are domestic airdromes with low air traffic intensity. Aerodrome (approach) NDBs are installed in the airdrome zone on the extended centerline of the runway, at a fixed distance from the runway threshold, and are intended to lead aircraft to the airdrome zone, to provide approach maneuvering and approach track keeping.

In Russia, such landing system is called Oborudovanie Sistemy Posadki (landing facilities) (OSP). The distance from the runway threshold is a criterion for distinguishing inner (called **near**) and outer (called **far**) NDBs in the OSP.

Marker beacons (MB) are installed together with the NDB in the OSP system, thus forming an integrated NDB-MB station. Marker beacons are also included in the ILS (Outer Marker (OM), Middle Marker (MM), Inner Marker (IM)).

A marker beacon is a transmitter operating at a fixed frequency of 75 MHz with an antenna having an antenna pattern lens (elliptical) shaped or bone shaped in horizontal plane view. Marker beacons are designed for identification (marking) of definite points on the earth surface and waypoints and operate in conjunction with aircraft marker radio receivers.

A diagram of MB and NDB placement by OSP standards is shown in Fig. 2.8 (a—double-marker placement, b—triple-marker placement used in case of challenging terrain near the runway threshold).

Diagram of MB and NDB placement by ICAO standards is shown in Fig. 2.9.

The diagrams of radio beacons placement differ in the distance between a runway threshold and an outer (far) NDB-MB (4,000 m by OSP and 7,200 m by ICAO standards).

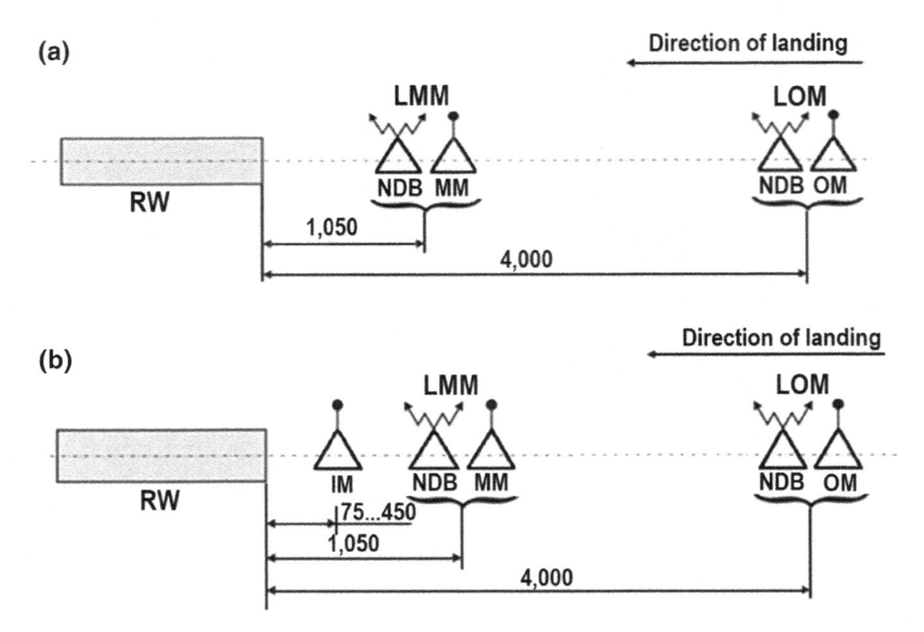

Fig. 2.8 Placement of radio beacons by OSP standards

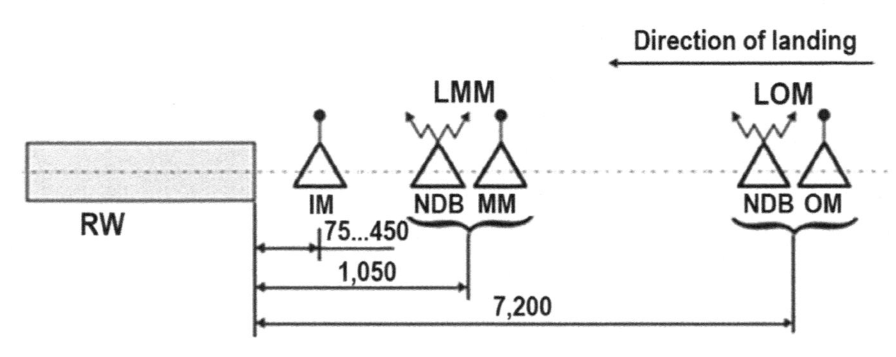

Fig. 2.9 Placement of radio beacons by ICAO standards

In Russia, an en route NDB and an outer (far) NDB use a two-letter call sign, while an inner (near) NDB uses a one-letter call sign (as a rule, the first letter of the outer (far) NDB call sign). In Europe and the USA, a three-letter call sign can be assigned to an en route NDB.

The NDBs can have different notations in aeronautical charts depending on their missions and chart manufacturers. Aeronautical charts typically label the NDBs with the symbols represented in Fig. 2.10 (a—NDB, b—NDB-MB, c—en route NDB).

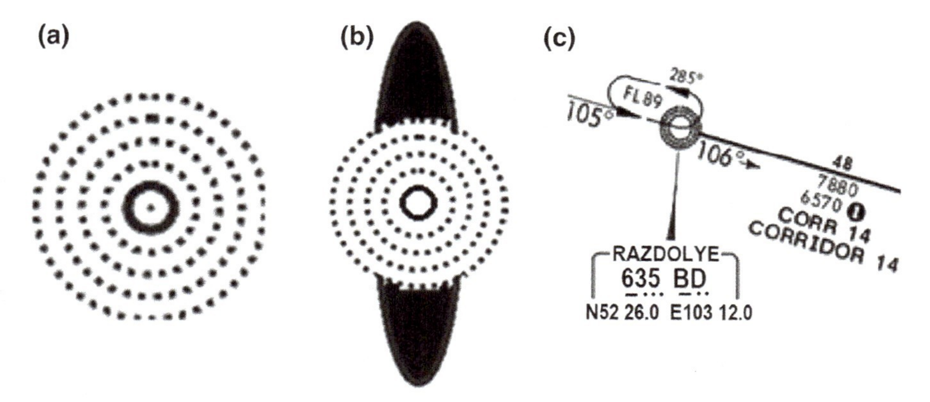

Fig. 2.10 NDB notation in aeronautical charts

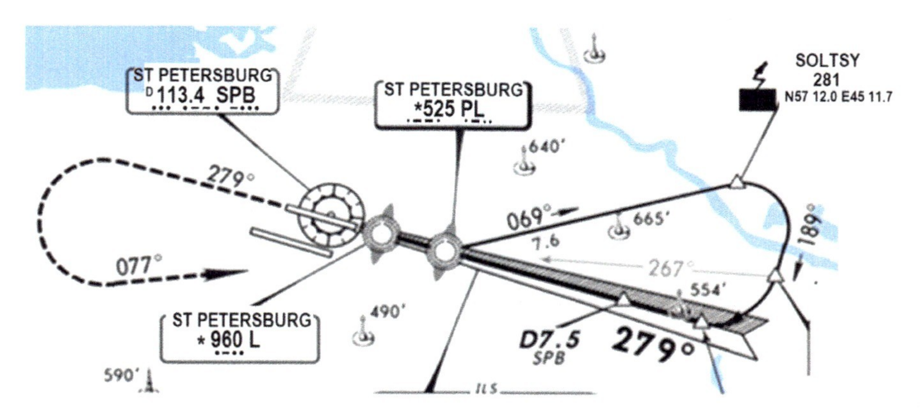

Fig. 2.11 LOM and LMM presentation in a navigational chart

In a chart, each radio navigational aid including an en route NDB has a "box" (Fig. 2.10c), a rectangle containing information identifying the aid. The data usually include:

name of the NDB (RAZDOLYE) corresponding to an inhabited locality or another control point;
a frequency (635 kHz);
a call sign (BD) which may be repeated in Morse code;
geodesic coordinates (north latitude N52°26.0′, eastern longitude E103°12.0′).

If a station does not operate round the clock, there is a "star" in front of the frequency value (ref. Fig. 2.11). The station's operating hours can be seen in the aeronautical information publications (AIPs).

Broadcasting radio stations are labeled with symbol in Russian charts (ref. Fig. 2.11). The "box" contains the name of a radio station, its frequency and

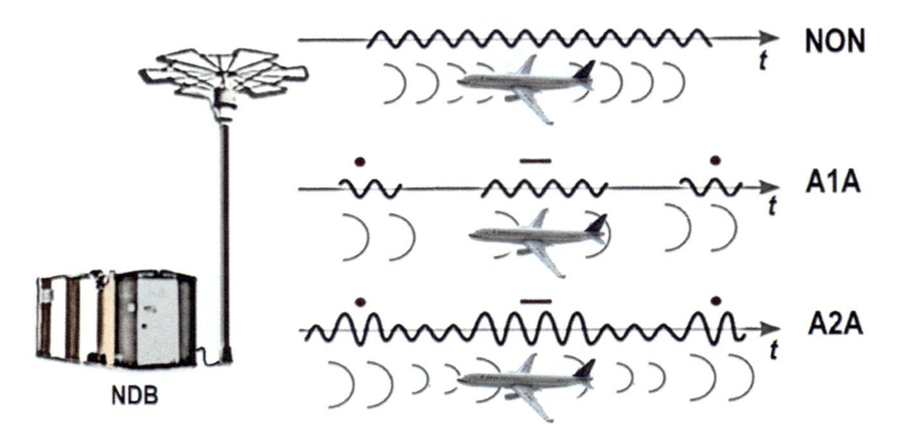

Fig. 2.12 NDB's emission

coordinates. There is obviously no call sign as the broadcasting radio stations do not transmit it.

2.2.3 A Generalized Block Diagram and Operation Principle of an NDB

The NDBs have three types of emission: N0N, A1A, and A2A (Fig. 2.12). Russian-made NDBs sometimes use A3E emission (double-band telephony, broadcasting) as well.

The N0N part of the emission is the transmission of an unmodulated carrier wave. While receiving the signal, an ADF determines and indicates the relative bearing of the NDB but call signs are not heard. To ensure that the carrier is received, a beat frequency oscillator (BFO) produces an offset frequency within the ADF receiver which when combined with the received frequency produces a signal with tone amplitude modulation 400 or 1,020 Hz frequency. It is the tone which is heard in a pilot's headphone set.

The A1A (amplitude telegraphy) part is the emission of an interrupted unmodulated carrier wave which requires the BFO to be on for aural reception. A2A (amplitude tone telegraphy) is the emission of an amplitude-modulated signal which can be heard on a normal receiver.

Hence, when using A1A and N0N beacons, the BFO switch should be selected ON for manual tuning, identification, and monitoring. A2A beacons require the BFO switch ON for manual tuning but OFF for NDB identification and monitoring. The BFO switch may be labeled "TONE" or "TONE/VOICE" on some ADF equipment.

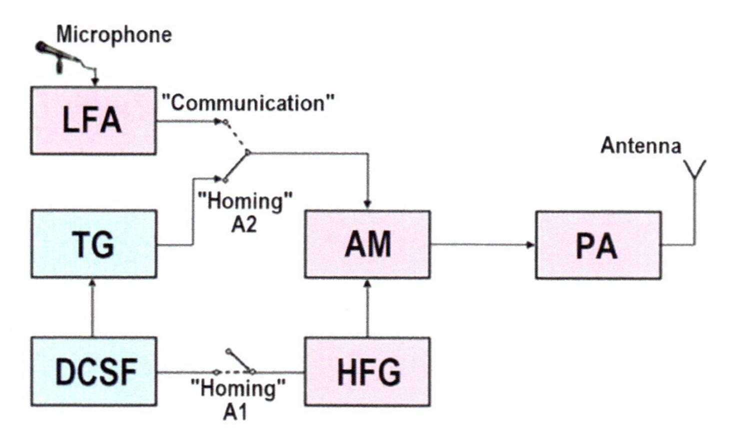

Fig. 2.13 Simplified block diagram of an NDB

BFO switch may be labeled “ТЛФ” (“Telephone”) or “ТЛГ” (“Telegraph”) on control panels of Russian-made ADFs.

NDBs can operate in the following modes:

“**Homing**”—telegraph electrical mode (A2A type of emission, modulation with frequencies 400 or 1,020 Hz), tone electrical mode (A1A type of emission). This is the primary mode of NDB operation.
“**Communication**”—is used for transferring information at the NDB frequency from a control tower to an aircraft if an onboard or ground VHF radio station fails. NDB identification signals are not transmitted in this mode. A “Translation” voice transmission mode is used.
In “**Translation**” voice transmission mode (A3E type of emission), voice information is transferred from a control tower to an aircraft by means of amplitude modulation of a carrier with a microphone output signal.
In outdated NDBs, “Telegraph,” “Tone,” “Microphone” modes were applied.
The “Telegraph” mode provided transferring information by means of key-controlled amplitude modulation of a carrier.
The “Tone” voice transmission mode provided transferring information by means of amplitude modulation of a carrier by key clicks.

The simplified block diagram of an NDB is shown in Fig. 2.13. The diagram represents: LFA—low-frequency amplifier; TG—tone generator-producing modulation signals with a frequency of 400 Hz or 1,020 Hz; DCSF—device for call sign formation (Morse code symbols); HFG—high-frequency generator (of a carrier); AM—amplitude modulator; PA—power amplifier.

Diagrams of output signals of NDB operational units are shown in Fig. 2.14.

In “Homing” mode (A2A type of emission), a voltage with an audio frequency of 1,020 Hz or 400 Hz is formed at the output of the tone generator TG.

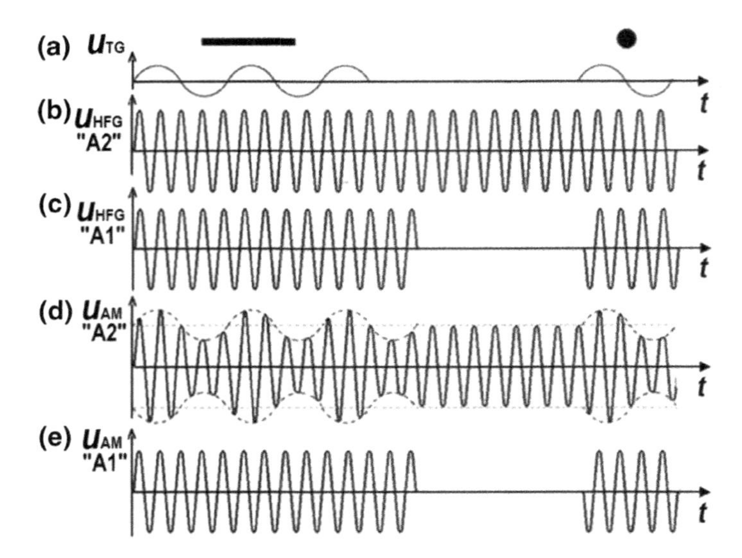

Fig. 2.14 Diagrams of output signals of NDB operational units

The automated device for call sign formation DCSF controls the TG operation by switching its power supply. The power is supplied for time segments corresponding to the duration of a Morse dash or dot assigned to the NDB. At the TG output, voltage impulses with an audio frequency are created which are as long as the corresponding Morse dashes or dots (Fig. 2.14a).

The HFG forms a continuous signal of carrier frequency (Fig. 2.14b, c). The amplitude modulator AM modulates HF oscillations of the high-frequency generator (HFG) by the TG voltage. The produced amplitude-modulated signal is amplified by the PA and then enters the antenna (Fig. 2.14d).

For higher NDB power (its bigger coverage), the A1A type of transmission is used (Fig. 2.14e).

In this submode, the DCSF controls the HFG operation so that at the HFG output, either HF oscillations are formed or there are no oscillations (discontinuity of the carrier). The signal passes through the AM to the PA, and at the PA output, tone-unmodulated oscillations with carrier discontinuity are formed.

As mentioned before, when an ADF receiver receives such oscillations, the call signs are not heard in the headphones. For their aural recognition, the ADF receiver applies its own inner amplitude modulation.

In “Communication” mode, the HFG signal is modulated with amplified LF voltage coming from a controller’s microphone. A controller’s instructions are heard in headphones connected to the ADF receiver output.

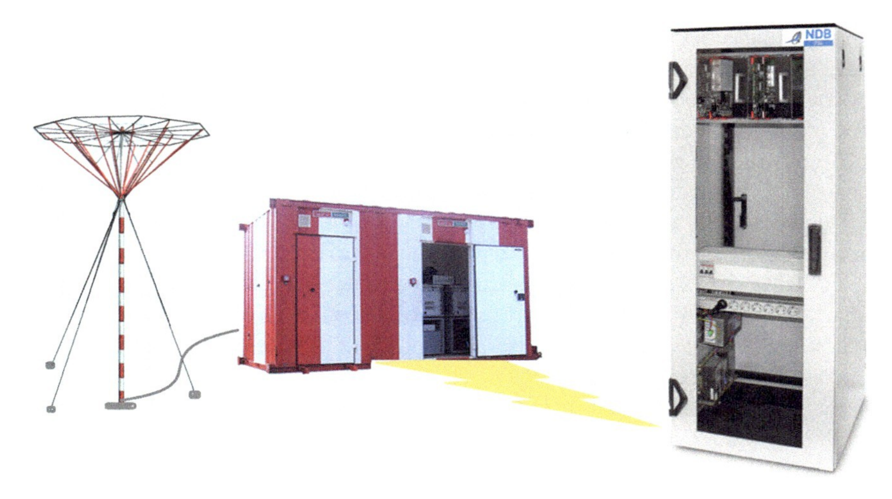

Fig. 2.15 Stationary variant of NDB placement. *Source* http://www.nports.ru/#!/catalog/ by permission of LLC Research and Development Company «Radio engineering systems»

Fig. 2.16 Mobile variant of NDB placement

2.2.4 NDB Design and Basic Specifications

NDBs differ in radiated power (high-, medium-, and low-powered) method of forming a carrier frequency and its stabilizing degree, level of test and control automatization, placement (a stationary or mobile variant).

High-powered NDBs are stationary, as a rule (Fig. 2.15).

Medium- and low-powered NDBs can be stationary or mobile (Fig. 2.16—Russian-made NDB PAR-10MA).

Equipment of a mobile NDB is housed in a car body. Antenna elements are attached to the car body during transportation. The mobile variant is usually used in military aviation.

A stationary variant can accommodate inner, outer, and en route NDBs. In this case, the NDB equipment (typically two sets) is installed in a special container (operating room) or in a building. The equipment of both sets is mounted in a common box. Marker beacons, if included, are mounted in the same box.

A construction difference of NDBs is in antenna size (antenna mast of an inner (near) NDB has a height of 5–6 m, that of an outer (far) NDB has a height of 20 m).

Design specifications and operation principle of an NDB will be discussed in more detail using the example of Russian-made RMP-200 equipment [2].

The RMP-200 radio beacon can be used as an outer (far), inner (near) single beacon or as a part of OSP system. Use of the radio beacon as an outer or inner beacon is determined by antenna size and radiation power adjustment when introducing the beacon into service.

Radio beacon equipment is located either in the operating room or in stationary buildings.

A typical (basic) radio beacon is composed of two sets of NDB equipment and two MB sets. However, the configuration can vary in presence/absence of a marker beacon as well as in the number of NDB and marker beacon sets.

The main peculiarities of the radio beacon RMP-200 are:

program-controlled digital formation of carrier HF oscillations, modulation frequencies, identification signals;
equipment of a NDB and a marker beacon have common power supply and control but are operationally independent;
continuous range verification of emitted signals;
built-in test check;
remote control, setting of main parameters, and radio beacon operation control.

Make time of an armed radio beacon is no more than 30 s. If the primary set fails, the standby set takes over automatically; transition time is no more than 2 s.

Transition to the standby set takes place if:

carrier power decreases below 50% in comparison with the level necessary for effective coverage;
an identification signal is not transmitted or does not correspond to the specified code;
an amplitude modulation index is half as large as its rating value;
a checking device fails.

The information panel provides data about the radio beacon status by means of light and audible indication. The following conditions are displayed:

shutdown of the failed equipment set and activation of the standby one;
code change or break of identification signal transmission;
decrease of carrier power by 20% in comparison with the level necessary for effective coverage;
decrease of an amplitude modulation index below 70%;
failure of the checking device.

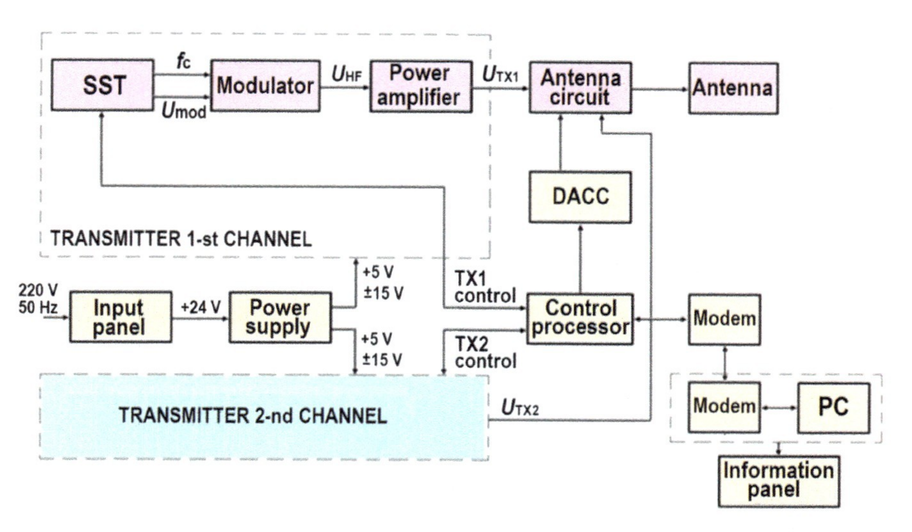

Fig. 2.17 Functional diagram of the RMP-200

Radio beacon RMP-200 control can be local or remote. The remote control is accomplished via a two-wire line or a radio line using a remote control unit and modems or radio modems at a distance of 0.5–10 km from the radio beacon.

The continuous presence of maintenance personnel is not required. The radio beacon equipment is meant for round-the-clock operation.

The radio beacon is composed of the following components:

an RMP-200 box, an input panel;
a NDB antenna (an AZ-20 umbrella-loaded vertical antenna or a T-type antenna with an installation height of 20 m for an outer NDB, an AZ-8 umbrella-loaded vertical antenna or a T-type antenna with an installation height of 5 m for an inner NDB);
an antenna of a marker beacon;
Air traffic control tower (ATCT) equipment comprising a personal computer with an application software package, a modem, and an information panel;
an operating room made of a container where equipment is installed for comfort operation of the RMP-200 box, input panel, and maintenance personnel. The input panel is located on the wall of the operating room (or an engineering building).

The primary parameters and specifications of the RMP-200 radio beacon are presented in Table 2.1.

The simplified functional diagram of the RMP-200 is represented in Fig. 2.17.

The control processor is intended to control the radio beacon in local and remote operation modes. The front panel of the control processor has status indicators (NORM, DEGRADATION, EMERGENCY) and control mode indicators (switching on/off the first or second equipment set, local/remote control mode). The front panel also contains mode controls of the radio beacon: buttons ON 1 SET, ON 2 SET, OFF 1 SET, OFF 2 SET, a toggle switch LC/RC.

Table 2.1 Primary parameters and specifications of the RMP-200 radio beacon

Parameter, specification	Value
Coverage area, at least (km)	
Outer Inner	150 50
Emission types	A1A, A2A, A3E
Minimal electric field intensity within the coverage (uV/m)	70
Working frequency band (kHz)	190-1,750
Formed frequency spacing (Hz)	100
Relative instability of a carrier frequency (%)	0,01
Frequency of a baseband identification signal (Hz)	400 ± 25 $1{,}020 \pm 50$
Identification signal	1-2-3 letters of Morse code
Speed of identification signal transmission	7 words/min
Repetition cycle of an identification signal (s)	
– A1A, A3E types of emission – A2A type of emission	10 60
Transmitter power, variable (W)	20–200
Amplitude modulation index in A2A type of emission (%)	85–100
Error of the radio beacon in measuring the relative bearing, σ, at most (°)	3

The signal shaper and tester (SST) consists of two functionally independent devices: a carrier frequency/modulation signal shaper and a tester of transmitter signals. The signal shaper operates in two modes:

generation of an identification Morse code signal;
amplification of a voice signal coming from the microphone.

Besides, the signal shaper generates the carrier frequency.

The identification signal generated by the signal shaper is a tone signal with a frequency of 400 or 1,020 Hz amplitude-shift keyed with Morse letters.

The shaper operation is monitored by a controller. The controller receives identification signal data from the control processor via data bus. The data contain a call sign code, a frequency of the modulating signal, an amplitude modulation index of the HF signal carrier, transmitter output power, a frequency of the HF signal carrier.

These parameters are recorded in non-volatile memory, thus enabling restart of signal generation after power loss by reading from memory without processor reset.

The controller forms codes for digital-to-analog converter (DAC) control for discrete intervals of time. The DAC forms an identification signal or a voice signal which is supplied to the signal shaper and tester output.

An HF signal of the carrier frequency is formed by a digital synthesizer in accordance with the carrier frequency control code coming from the controller.

A low-pass filter smooth step voltage coming from the digital frequency synthesizer and in this fashion the modulation voltage is formed.

The tester is designed to check parameters of transmitter signals: an amplitude modulation index, power of signals coming from the transmitter to the antenna, a frequency of the HF signal carrier, a frequency of a modulation signal, a harmonic factor of the HF signal envelope, current consumed by the power amplifier, temperature of the power amplifier, correspondence of the identification code to the specified value.

The modulator provides amplification and amplitude modulation with an audio frequency of the HF signal coming from the signal shaper and tester (SST).

The power amplifier amplifies signals coming from the modulator.

The device for antenna circuit control (DACC) is intended to maintain the maximal current of antenna when antenna parameters change. It consists of three operationally independent devices: a control circuit of the HF switch, a voice signal amplifier, a device for self-tuning of the antenna circuit.

The antenna circuit is designed for coordinating the transmitters and the antenna as well as for connecting the operating transmitter set to the antenna.

The antenna is a shortened vertical radiator fed relative to the ground plane.

The input panel automatically connects the main or standby network to the RMP-200 equipment and supplies the voltage for lighting and electrical outlets.

The information panel is intended to display the status of two radio beacons (it provides independent status signaling of two NDB and two marker beacons). It has a wire link with the ATCT computer which supplies the information about the status of the radio beacons ("OPERATION," "DEGRADATION," or "EMERGENCY"). The "EMERGENCY" indicator lights up if both radio beacon sets fail. The "DEGRADATION" indicator lights up if one of the radio beacon sets fails. If any of four radio beacons fails, the controller generates a discrete audio signal which can be turned off manually with a toggle switch "TONE OFF."

2.3 Aviation VHF Direction Finders

2.3.1 Function and General Characteristics

Radio direction finding is measuring the direction (bearing) of a radio-frequency radiation source (a radio station) with the aid of radio facilities (direction finders).

The bearing information is formed by the direction finder antenna system due to its moving (rotating) or moving its antenna pattern. There are also antenna systems with fixed, specially oriented antenna patterns (e.g., in northsouth and westeast directions). The antenna systems of direction finders are designed to receive radio waves with vertical polarization.

The antenna system transforms the direction of a radiation source into such parameters of an output radio signal as amplitude, frequency, phase, reception moment. The bearing information can also be contained in the parameters of output signal modulation: in modulation depth or envelope phase under the amplitude modulation; in modulation parameters under the frequency (phase) modulation.

Processing the signal in a DF enables to extract the bearing information from the radio signal parameters or from the parameters of its modulation.

The true bearing is reckoned from the north direction of geographical meridian taken as a reference point to the direction. The direction finder, therefore, creates a reference signal whose parameter sets the reference point.

Modern aviation direction finders (DFs) are intended to control the aircraft attitude in the air by direction finding of signals from onboard radio stations of VHF band (118–137 MHz, class A3 emission) and UHF band (220–400 MHz, military airplanes) during air-to-ground and air-to-air communication. The DF provides steady direction finding of onboard radio stations if the transmission is not shorter than 0.5 s.

DFs used for direction finding of VHF radio stations are also designated with VDF abbreviation.

The DF enables to determine the true bearing of an aircraft, i.e., the angle between the north reference direction of geographical meridian passing through the DF antenna center and the aircraft direction (Fig. 2.18). The measured bearing is displayed either on the DF equipment indicator or on a controller's indicator. The term "azimuth" sometimes is used instead of "true bearing."

DFs are widely used in the ATC system to control the aircraft movements on air tracks (en route DF) and terminal areas (aerodrome DF). If there are several DFs, linked to a network, the aircraft's position can be fixed using triangulation or LOP method.

Aerodrome DFs intended for landing, circuit, and approach air communication channels are located near the surveillance radar (approach radar system). If the aerodrome is not equipped with ILS, the DF is located along an extension of the runway centerline up to 500 m apart from the outer NDB (Fig. 2.19).

DFs intended for air communication channels of ATC centers can be located in the vicinity of en route surveillance radars.

DF installation next to the aerodrome or en route surveillance radars must meet electromagnetic compatibility requirements.

DF range is determined by the flying height of an aircraft and is practically independent on the emissive power of an onboard radio station since VHF radio waves propagate within the line-of-sight distance. For example, at the flying height of 10,000 m the direction-finding range is about 350–400 km.

The DF antenna system has a limited sector of signal reception in the vertical plane (from the earth surface up to 45–60°). Accordingly, above the DF antenna there is a "dead area" ("cone of ambiguity") with the radius equivalent to the flying height of an aircraft.

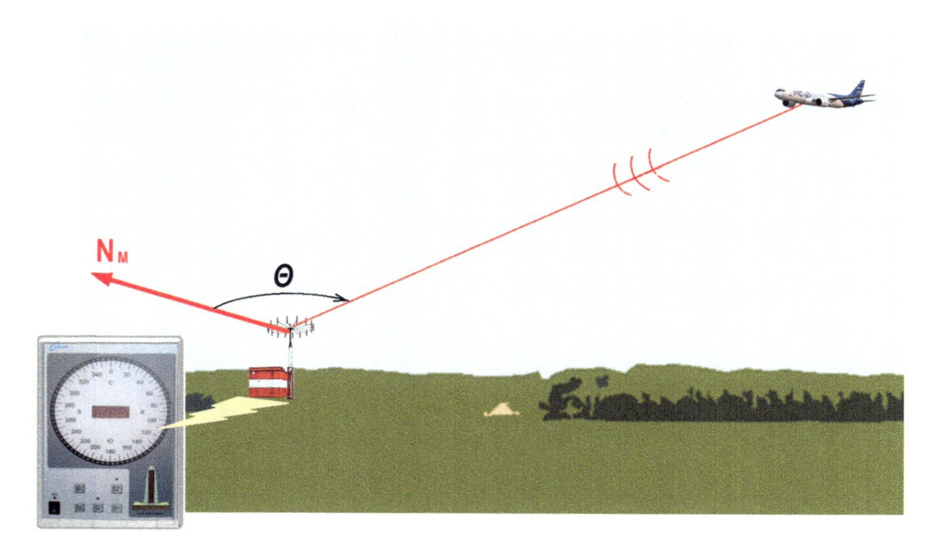

Fig. 2.18 Principle of direction finding of an aircraft radio station

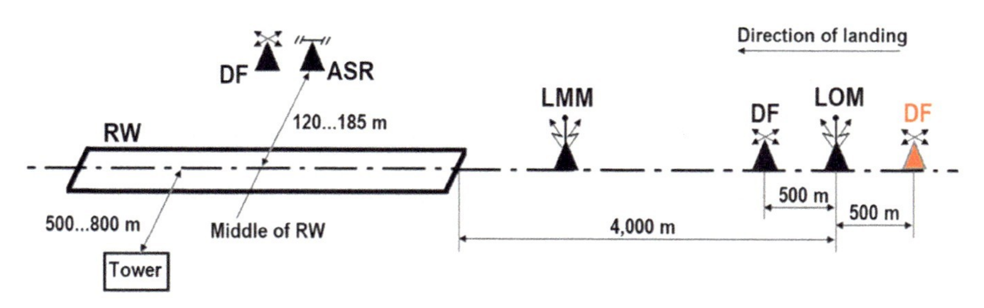

Fig. 2.19 DF siting at the aerodrome

An important characteristic of a DF is the number of frequency channels for simultaneous operation. ATC controllers employ different areas of responsibility and, correspondingly, different frequencies for communication with crews. To provide continuous monitoring of the aircraft attitude on approach and landing, the airdrome DF must have several (up to 16) DF channels working simultaneously.

The DFs were the first navigation systems and date back to the very beginning of the twentieth century. Over the years, considerable changes have taken place in their techniques of forming and processing information as well as in antenna system design.

The direction-finding accuracy greatly depends upon the properties of DF antenna systems. As a consequence, there are a number of specific and rather tough requirements for an antenna system. It must:

have a high steepness and angular sensitivity of the antenna polar diagram in usable sections so that slight bearing changes cause significant increments of output signals;
have a highly stable antenna polar diagram, i.e., preserve its form and attitude with the changes of properties of soil and environment, as well as under the action of re-radiation of surrounding terrain features;
possess a bandpass response, i.e., the antenna polar diagram must be uniform over the whole frequency range of usable radio waves;
have minimal dimensions and weight as some types of DFs need rotating or must be placed on limited areas.

According to the antenna system design, the following variants of DFs are distinguished:

with horizontal mechanical rotation of the antenna system (in the form of a loop antenna or a pair of anti-phased vertical dipoles), whose antenna pattern has a distinct minimum (Fig. 2.20a);
with a fixed antenna system containing two pairs of anti-phased vertical dipoles oriented in north–south and west–east directions and a central non-directional antenna (Fig. 2.20b);
with a fixed antenna system containing two (as Fig. 2.20b demonstrated) or more pairs of anti-phased vertical dipoles and a central non-directional antenna when the bearing information is formed by adding antenna signals. The summarized antenna pattern thus rotates electronically;
with rotation of a non-directional antenna (or switching dipoles spaced circumferentially) around the central (also non-directional) antenna. It is the method that is used in modern Doppler DFs (Fig. 2.20d).
with rotation of a non-directional antennas pair (or pair switching neighboring dipoles spaced circumferentially). The method implies measuring the signal-phase difference between two adjacent switched dipoles (Fig. 2.20c). It is the method that is used in RDF-734 (manufactured by LLC Research and Development Company «Radio engineering systems», Russia).

According to the methods of forming and processing a signal, the following types of aviation DFs are employed:

amplitude DFs, in which the bearing information is contained in signal modulation or in the amplitude modulation factor of a signal formed at the antenna system output (these were ARP-6, ARP-7 in the USSR/Russia);
amplitude-phase DFs, in which the bearing information is contained in the envelope phase shift of an amplitude-modulated signal (these were ARP-10, ARP-11 in the USSR/Russia);
Doppler DFs (modern), in which the bearing information is contained in Doppler frequency shift of a signal received by a fixed antenna moving relative to the radiation source (these are ARP-75, ARP-80, ARP-95, DF-2000 in Russia).
differential Doppler DFs (modern) in which the bearing information is contained in signal-phase difference between two adjacent switched dipoles (RDF-734, Russia).

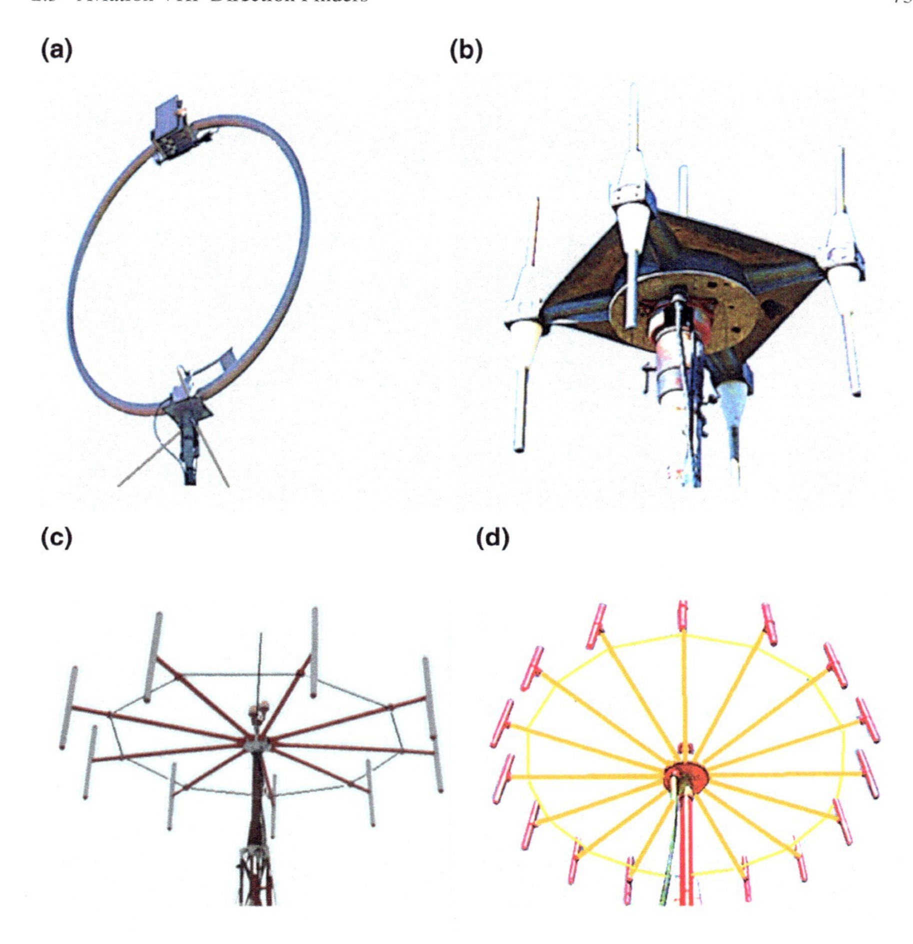

Fig. 2.20 Types of DF antennas

The bearing can also be determined with the use of other signal parameters. So, the bearing information can be extracted from the phase differential of signals received by mutually spaced non-directional antennas (phase DFs or interferometers) or from the difference of their reception moments also of non-directional antennas. However, such types of DFs have not found any application in aviation.

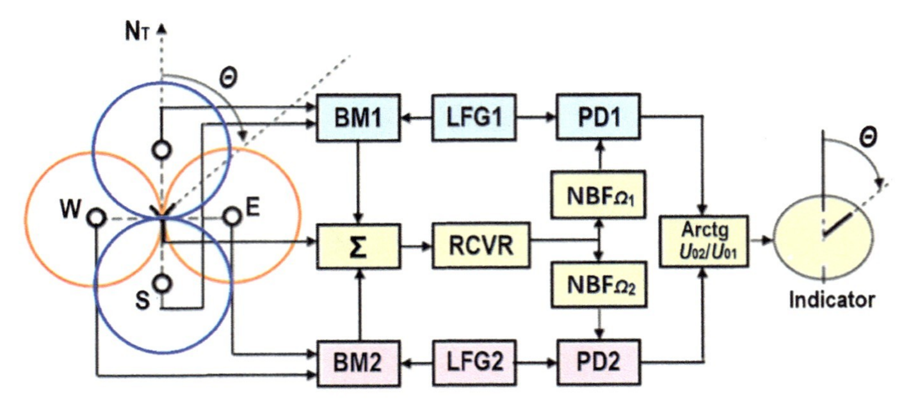

Fig. 2.21 Simplified block diagram of an amplitude DF

2.3.2 Principles of Forming and Processing Information in DFs

Amplitude DFs. The aviation amplitude DFs are based on the dependence of amplitude of received signals on aircraft bearing. This dependence occurs due to the antenna polar diagram and when processing can be transformed into the dependence of amplitude modulation parameters on bearing by combining signals from antennas outputs.

The antenna system of amplitude DFs consists of two mutually perpendicular pairs of dipoles oriented in north–south and west–east directions and a central non-directional antenna.

Such antennas possess good bandpass response and form horizontal eight-shaped antenna patterns F_1 (Θ) and F_2 (Θ) (Fig. 2.21) with two directions of zero reception shifted on 180°. Properties of such antenna pattern underpin the **method of minimum**.

Signals received from antenna pairs have the same frequency and are jointly processed (amplified) in the common receive path. For the following separation, their local amplitude modulation with different low frequencies Ω_1 and Ω_2 is used. Meanwhile, modulation depths m_1 and m_2 are proportional to the amplitudes of signals received by the corresponding antennas which, in their turn, depend on the bearing of the radio station.

Thus, in the DF, the dependence of amplitude of received signals on the direction of incoming radio waves is transformed into the dependence of depth of amplitude modulation of the same signals. The direction of the radio station is determined by comparing the depths of local modulation of signals received by the pairs of dipoles.

The simplified block diagram of a DF is shown in Fig. 2.21.

A signal from the aircraft radio station which is located in direction Θ induces the emf in the directional antennas which can be defined as:

$$e_{N-S} = E_{md} \cos \Theta \sin \omega t$$

$$e_{E-W} = E_{md} \sin \Theta \sin \omega t,$$

where E_{md} is an amplitude of antenna emf in the direction of maximal reception.

Output signals from each pair of dipoles are supplied to the input of a corresponding balanced modulator (BM1 and BM2) where the VHF signal is modulated by the signal of a respective low-frequency generator (LFG1 and LFG2) with frequencies Ω_1 and Ω_2. As a result, the signals of modulation side band only (SBO) are formed at the outputs of BM1 and BM2 balanced modulators as:

$$u_{N-S} = U_{md} \cos \Theta \sin \Omega_1 t \sin \omega t$$

$$u_{E-W} = U_{md} \sin \Theta \sin \Omega_2 t \sin \omega t.$$

These signals are summarized with signal u_a of the non-directional central antenna and are supplied to the input of a common receiver (RCVR). Phasing the signal of the non-directional antenna is necessary because the phase shift between output signals of the central and directional antennas is equal to $\pi/2$. The signal is phased in an antenna amplifier as a part of a summator. The signal of the central antenna after phasing is written as:

$$u_a = U_{ma} \sin \omega t.$$

From the summator output to the receiver input, a signal is supplied:

$$u_\Sigma = U_{ma}\left(1 + \frac{U_{md}}{U_{ma}} \cos \Theta \sin \Omega_1 t + \frac{U_{md}}{U_{ma}} \sin \Theta \sin \Omega_2 t\right) \sin \omega t. \tag{2.1}$$

Thus, the signal of (2.1) type is amplitude-modulated by frequencies Ω_1 and Ω_2 with modulation depths $m_1 = \frac{U_{md}}{U_{ma}} \cos \Theta$ and $m_2 = \frac{U_{md}}{U_{ma}} \sin \Theta$ dependent on bearing of the radio station. To exclude the over-modulation which causes a bearing error, condition $m_1 + m_2 \leq 1$ needs to be met.

The output signal of a receiver after detecting is the sum of two variable signals with frequencies Ω_1 and Ω_2 as:

$$u_{out} = U_{md} \cos \Theta \sin \Omega_1 t + U_{md} \sin \Theta \sin \Omega_2 t. \tag{2.2}$$

The signal (2.2) is supplied to narrow-band filters $NBF_{\Omega 1}$ and $NBF_{\Omega 2}$ tuned to frequencies Ω_1 and Ω_2. At the filter outputs, the varying voltage of frequencies Ω_1 and Ω_2 is formed and supplied to the inputs of the phase discriminators PD1 and PD2, correspondingly. The other inputs of the phase discriminators receive reference signals from LFG1 and LFG2 with frequencies Ω_1 and Ω_2, correspondingly.

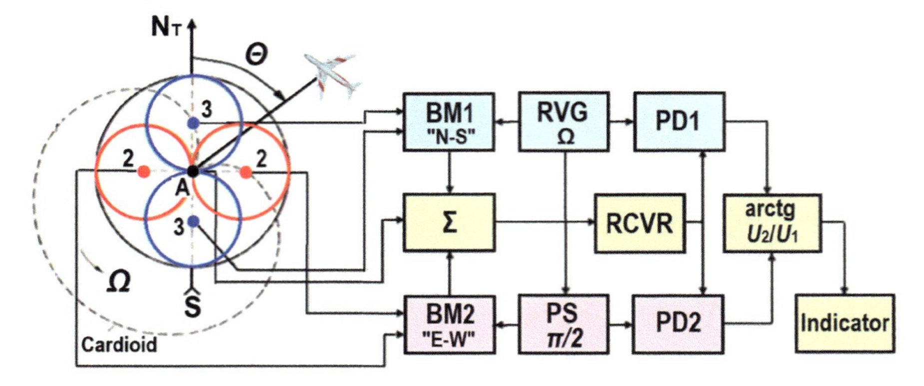

Fig. 2.22 Simplified block diagram of an amplitude-phase DF

At the outputs of PDs, the fixed voltage is formed as:

$$U_{01} = K_1 U_{md} \cos \Theta$$
$$U_{02} = K_2 U_{md} \sin \Theta,$$

where K_1, K_2 are transmission gains of the PDs.

Comparison of the PD output voltage values enables to calculate bearing of the radio station $\Theta = \text{arctg}(U_{02}/U_{01})$.

The main advantage of this DF is the simplicity of its schematic and a common reception and amplification path.

Its disadvantages are:

low interference immunity against external disturbances;
lack of the bearing resolution ability;
inability of direction finding by time-limited signals (duration of a received signal must be many times longer than the period of modulation frequencies). Meanwhile, the value of the lower modulation frequency must be larger than the higher frequency of the voice spectrum (3 kHz) as onboard radio-bearing stations are specifically designed for transmitting voice signals.

Amplitude-phase DFs. The amplitude-phase method is a combined one. Here, as in the amplitude method, directional antennas are used but they have antenna patterns rotating in the horizontal plane. As a result of rotating the antenna patterns (mechanically or electronically), the bearing information is contained in the envelope phase of an amplitude-modulated signal which is formed on adding the antenna signals.

A simplified block diagram of an amplitude-phase DF with electrically scanned antenna pattern is shown in Fig. 2.22.

The antenna system of a radio beacon consists of three antennas: a non-directional central antenna A1 and two antennas A2 and A3 with mutually perpendicular bases, each comprising two anti-phased vertical dipoles. Antennas A2 and A3 are oriented in west–east and north–south directions, correspondingly, and have eight-shaped antenna patterns, i.e.,

$$F_2(\Theta) = \sin(\Theta),\ F_3(\Theta) = \cos(\Theta).$$

The antenna pattern rotates electronically in the DF itself due to appropriate processing of signals coming from the antenna system.

The output signal of the central antenna A1 is represented as:

$$u_{\mathrm{A1}} = U_{\mathrm{A1m}} \sin \omega t,$$

where $U_{\mathrm{A1}m}$ is the amplitude of a signal.

Output signals of the directional antennas A2 and A3 after their balance modulation by signals phase-shifted by $\pi/2$, having frequency Ω and coming from the low-frequency (LF) reference voltage generator (RVG), are represented as:

$$u_{\mathrm{A2}} = U_{\mathrm{A2m}} \sin \Theta \sin \Omega t \sin \omega t$$

$$u_{\mathrm{A3}} = U_{\mathrm{A3}m} \cos \Theta \cos \Omega t \sin \omega t,$$

where U_{A2m}, $\mathrm{U_{A3m}}$ are the amplitudes of output signals of the directional antennas.

Let us assume that the amplitudes of output signals are equal, i.e., $U_{\mathrm{A1m}} = U_{\mathrm{A2m}} = U_{\mathrm{A3m}} = U_{\mathrm{m}}$. Then, in their addition, the signal entering the DF receiver (RCVR) input can be written as:

$$\begin{aligned} u_{\Sigma} &= U_m(1 + m \sin \Theta \sin \Omega t + m \cos \Theta \cos \Omega t)(1+) \sin \omega t \\ &= U_m[1 + m \cos(\Omega t - \Theta)] \sin \omega t. \end{aligned} \tag{2.3}$$

Such a signal is formed at the output of the antenna system with a cardioid antenna pattern (Fig. 2.22) rotating in the horizontal plane with the angular frequency Ω.

From Eq. (2.3), it is clear that the bearing Θ information is contained in the envelope phase of the summarized amplitude-modulated signal. Meanwhile, the envelope phase of the summarized amplitude-modulated signal with frequency Ω constantly changes, being equivalent to mechanical rotation of the directional antennas with the same frequency.

After reception and detection of the summarized RCVR output signal, a LF signal of variable phase is formed as:

$$U_s = U_m \cos(\Omega t - \Theta),$$

whose phase contains the bearing Θ information and depends on the current value of the RVG signal phase.

The signal enters the phase discriminators PD1 and PD2. The RVG supplies reference voltage to the other PD inputs (voltage directly from the RVG is supplied to PD1, voltage phase-shifted (PS) by $\pi/2$ is supplied to PD2). At the PD outputs, the fixed voltage is formed. It is proportional to the cosine of phase difference of input voltage, i.e.,

$$U_1 = K_1 U_m \cos \Theta$$

$$U_2 = K_2 U_m \sin \Theta,$$

where K_1 and K_2 are transmission gains of PDs. If K_1 is equal to K_2, the bearing is determined by a computing device according to the equation:

$$\Theta = arctg(U_2/U_1).$$

The considered method of direction finding is used in ARP-10 and ARP-11 radio direction finders (USSR/Russia). The antenna pattern rotates mechanically in ARP-10 and electronically in ARP-11. For increased accuracy, ARP-11 has a more complicated antenna system consisting of four pairs of dipoles with bases mutually spaced at 45°.

Application of electronic rotation of antenna pattern increases the system reliability and enables to use much higher rotation frequency Ω, and hence, to increase the speed of space scanning. However, an accurate balancing of phase-meters and balance modulators is required since their non-identity creates an additional instrumental error of measurements.

Doppler DFs. The operation principle of Doppler DFs is based on Doppler-shift effect of the signal received by an antenna moving relative to the radiation source. Doppler frequency shift causes phase modulation of the received signal, its phase depending on the direction of the radiation source.

In its simplest form, the antenna system of a DF consists of two non-directional antennas—a central and a side one. The central antenna A2 is fixed. The side antenna A1 rotates in the horizontal plane along the circumference with radius r and with the constant angular velocity Ω_r around the central antenna (Fig. 2.23).

As antenna A1 at one moment approaches the radiation source and at the next moves away from it, Doppler effect arises and causes phase-spatial modulation of the received signal.

In the fixed antenna A2, emf is induced:

$$e_{A2} = E_{ma} \sin \omega_0 t.$$

The signal phase in antenna A1 differs from the signal phase in the central antenna by the value:

$$\Delta\varphi = 2\pi S/\lambda,$$

where S is the difference of signal path from antenna A1 to antenna A2 which is defined by the wave phase front; λ is the wavelength of received oscillations.

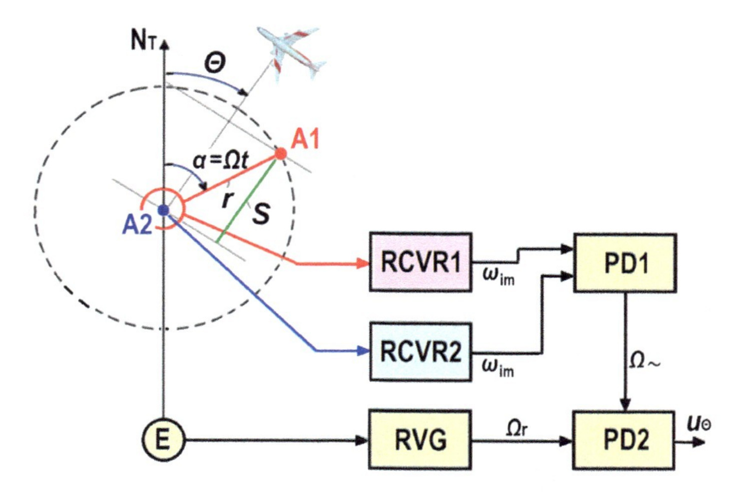

Fig. 2.23 Simplified block diagram of a Doppler DF

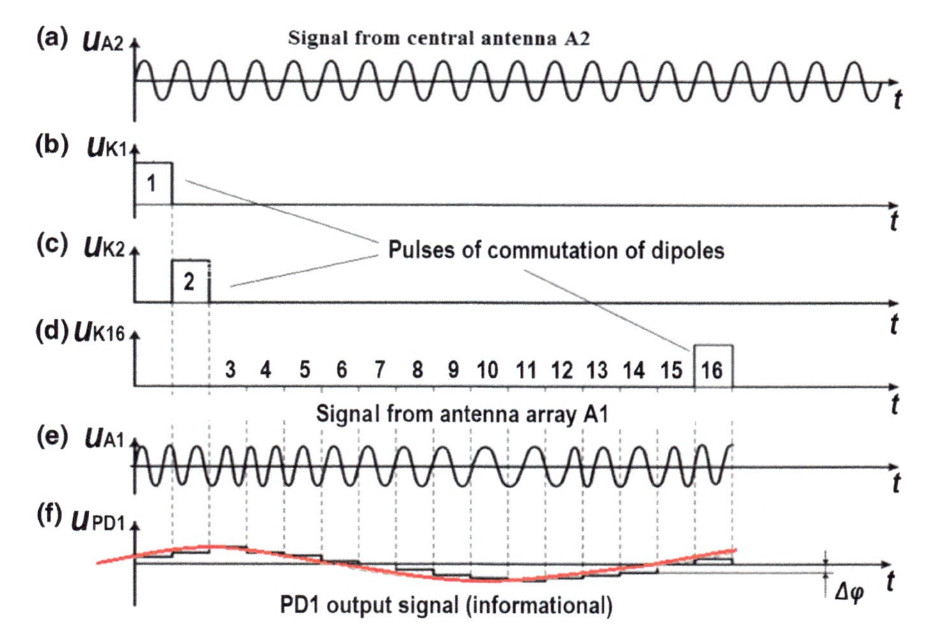

Fig. 2.24 Signal diagrams of a pseudo-Doppler DF

As it is shown in Fig. 2.24:

$$S = r\cos(\Omega_r t - \Theta),$$

where $\Omega_r t$ is the current angular position of the mobile antenna, and Θ is bearing of the radio station.

Consequently:

$$\Delta\varphi = \frac{2\pi r}{\lambda}\cos(\Omega_r t - \Theta).$$

Then in the rotating antenna, emf is induced:

$$e_{A1} = E_m \sin\left[\omega_0 t + m_\varphi \cos(\Omega_r t - \Theta)\right],$$

where $m_\varphi = 2\pi r/\lambda$ is the phase deviation.

Thus, the bearing information is contained in the phase of oscillation received by the mobile antenna (information signal). The information signal:

$$u_\Theta = K_{PD} m_\varphi \cos(\Omega_r t - \Theta), \tag{2.4}$$

can be formed by PD1 phase discriminator after comparing the signals of the mobile and central antennas (KPD is the PD transmission gain).

The bearing information can be obtained by comparing the phases of information and reference signals of type $u_{\text{ref}} = U_{m\,\text{ref}} \cos \Omega_r t$ in PD2.

The reference signal with frequency Ω_r corresponding to the rotation frequency of antenna A1 is generated by RVG. Its phase correlates with the current position of the antenna and is equal to zero when the antenna is directed to the north.

Enlargement of the relative size of base r/λ results in increase of DF angular sensitivity as the phase shift $\Delta\varphi$ increases with the constant bearing. Moreover, Doppler DFs have increased immunity against the disturbances caused by re-reflection of signals from the objects in DF coverage.

However, when radius r is large, the mechanical rotation of the antenna becomes a challenge. For that reason, a rotatable antenna is replaced by an antenna array composed of vertical dipoles (as a rule, sixteen) spaced circumferentially with the same radius r. The dipoles are alternately (Fig. 2.24b–d) switched to the receiver input by means of a mechanical or electronic switchboard and simulate antenna rotation.

The switched dipoles provide the change of the permanent modulation function for its discrete values (Fig. 2.24e) taken in the certain points of space. In this case, the pseudo-Doppler effect is that the phase of signals taken from the inputs of dipoles 1–16 changes discretely from dipole to dipole for a value $\Delta\varphi$ which is determined by the difference of travel path between adjacent dipoles. The value depends upon the number of dipoles and the line of travel of the electromagnetic wave. The cumulative curve of the phase change in one complete switching cycle composed of samples fits in Doppler envelope (Fig. 2.24f). Duration of a sample is time of reception by a specified dipole.

The further processing with the purpose to define the bearing includes the extraction of an envelope from the step function of phase shift (sine curve).

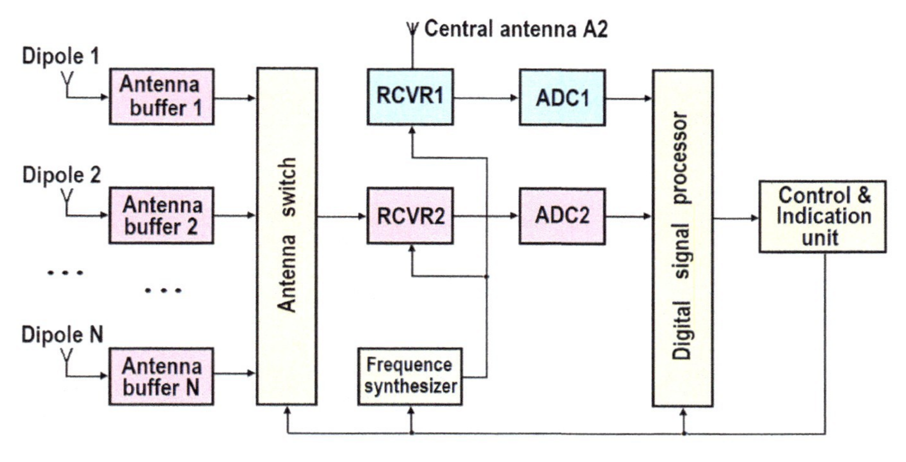

Fig. 2.25 Block diagram of a pseudo-Doppler DF

Frequency of discretization, in our case, of the distance between adjacent dipoles, is selected on the base of the Nyquist theorem. Provided that the maximal frequency of information signal spectrum $f_{\max}$ is equal to $m_\varphi \Omega_r / 4\pi^2$, the Nyquist theorem enables to show that the linear distance d between dipoles must meet the condition $d \leq \lambda/2$.

It should also be noted that the relative base size r/λ can be increased only to a limited extent, otherwise ambiguity of direction finding occurs which is typical for any phase method of measurements.

A simplified block diagram of a pseudo-Doppler DF with switched dipoles is shown in Fig. 2.25.

A diagram peculiarity is in the separate processing of output signals of the central antenna and the antenna array in receivers RCVR1 and RCVR2. LF output signals of the receivers get digitized (ADC1 and ADC2) after which the digital signal processor forms the bearing information.

The separate processing of antenna signals and use of two receivers is a significant drawback of the diagram. In practice, to simplify the DF block diagram, a common receiver is used for processing the signals of antennas A1 and A2.

To avoid signal overlapping in the common receiver, the VHF signal of antenna A1 enters the single-band modulator (SBM) for frequency translation $f_c + f_{\text{ref}}$ (Fig. 2.26). The signal of the central antenna A2 is not transformed. The frequency separation of signals from antennas A1 and A2 by the value f_{ref} is chosen so that both signals fall within the receiver passband. When adding the signals from antennas A1 and A2, a beat signal with the phase-shift information is formed in the common path. The further processing in the DF involves extraction of the information content from a common signal and then the digital processing provides formation of bearing information.

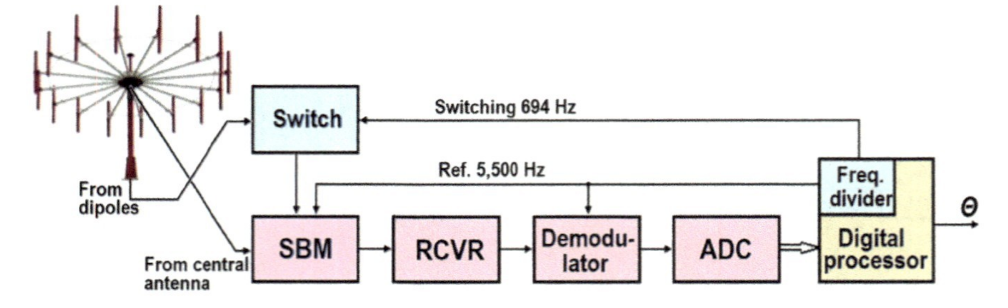

Fig. 2.26 Block diagram of a pseudo-Doppler DF with single-band modulator

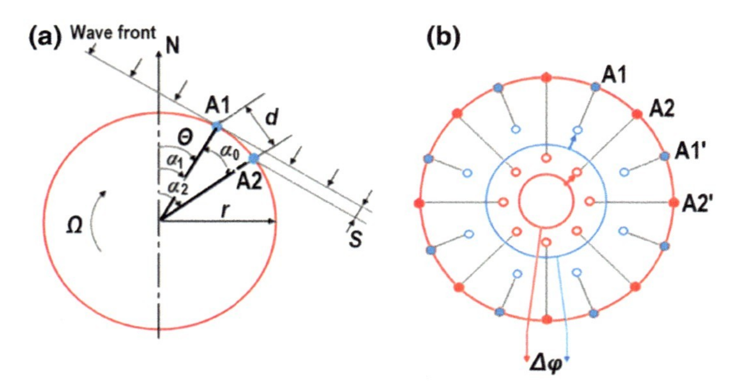

Fig. 2.27 Antenna system of a differential Doppler DF

This design concept based on the use of a single-band modulator and a common receive path is implemented in all Russian pseudo-Doppler FDs (ARP-75, ARP-80, ARP-95, DF-2000).

Differential Doppler DFs. To eliminate ambiguity of direction finding when increasing the relative size of base r/λ a differential Doppler method is applied.

The method implies measuring the phase difference either between emf induced in two non-directional antennas rotating simultaneously with radius r, or between two adjacent switched dipoles. The antenna system of such DF is shown in Fig. 2.27.

The instantaneous value of phase difference of output signals of antennas A1 и A2 whose position is characterized by angles α_1 and α_2 (where $\alpha_2 = \alpha_1 + \alpha_0$) is determined by the equation:

$$\begin{aligned}\Delta\varphi &= \frac{2\pi r}{\lambda}\sin(\alpha_2 - \Theta) - \frac{2\pi r}{\lambda}\sin(\alpha_1 - \Theta) \\ &= \frac{4\pi r}{\lambda}\cos\left(\frac{\alpha_2 + \alpha_1}{2} - \Theta\right)\sin\frac{\alpha_2 - \alpha_1}{2}.\end{aligned}$$

As the linear distance between adjacent antennas is:

$$d = 2r \sin \frac{\alpha_0}{2} = 2r \sin \frac{\alpha_2 - \alpha_1}{2},$$

the phase difference is defined as:

$$\Delta\varphi = \frac{2\pi d}{\lambda} \cos\left(\frac{\alpha_2 + \alpha_1}{2} - \Theta\right) \tag{2.5}$$

In the process of rotation (switching) of antennas with the angular velocity Ω, the current values of their turning angles are determined by the equations $\alpha_1(t) = \Omega t$ and $\alpha_2(t) = \Omega t + \alpha 0$. Plugging these equations in (2.5) gives:

$$\Delta\varphi = \frac{2\pi d}{\lambda} \cos\left(\Omega_r t + \tfrac{\alpha_0}{2} - \Theta\right) = m_\varphi \cos\left(\Omega_r t + \tfrac{\alpha_0}{2} - \Theta\right), \tag{2.6}$$

where $m_\varphi = 2\pi d/\lambda$ is phase deviation.

Thus, the phase difference of output signals of rotating antennas contains bearing information. To extract the information, it is necessary to have separate receivers for processing VHF signals from the aircraft radio station and forming a LF information signal with frequency Ω_r. Besides, there must be a reference signal synchronized with frequency Ω_r of antenna rotation (switching), signal phase $\Omega_r t$ carrying information about the current position of antennas in the process of rotation.

From Eq. (2.6), it is clear that the phase difference of signals of antennas A1 and A2 varies from zero to a maximal value equal to $\Delta_\varphi = 2\pi d/\lambda$. As the maximal value of the uniquely determined phase difference is equal to 2π, condition $d \leq \lambda/2$ must be met for unambiguous direction finding.

In practice, the distance between antennas is much smaller than the radius of antenna rotation and the phase deviation is smaller than that of Doppler DFs with one rotating antenna. It is possible therefore to reckon the bearing unambiguously with the use of large-based antennas. This provides high accuracy, angular sensitivity, and interference immunity against environmental reflections.

The considered principle of direction finding is implemented in DF-734 (NPO-RTS, Russia).

2.3.3 Design and Main Specifications of DFs

DFs differ in operating principle which defines the antenna system design as well as diagrams of the receive path and the bearing signal formation path. In practice, the amplitude-phase DFs, Doppler DFs (most widely used), and differential Doppler DFs are applied.

Amplitude-phase DFs have a small-sized antenna system so their mobile use is possible (Fig. 2.28a)—an issue of particular concern for military aviation. The antenna systems of Doppler and differential Doppler DFs have a diameter of about

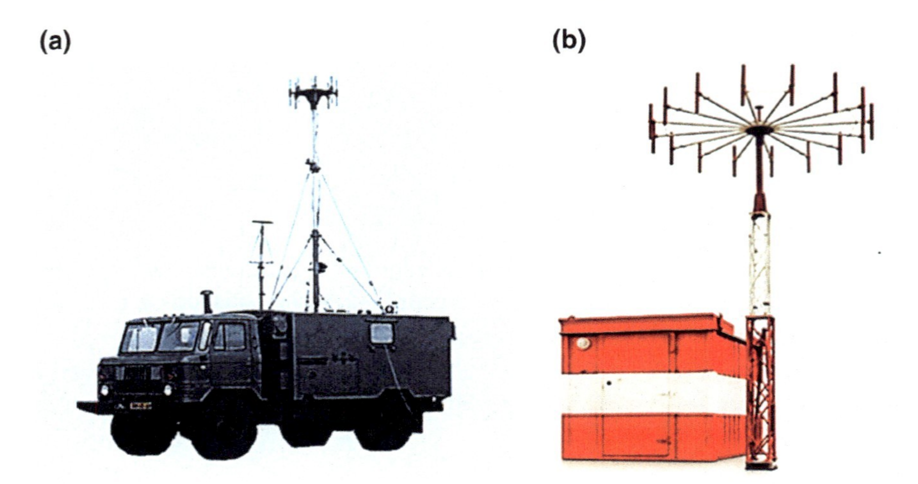

Fig. 2.28 Variants of DF placement

Table 2.2 DF performance specifications

Parameter	Value
Frequency band (MHz)	118–137
DF indicator error of direction finding, σ, at most (°):	
Any DFs Doppler DFs	2.5 1.5
Coverage, at least (km):	
At flying height of aircraft 1000 m At flying height of aircraft 3000 m	80 150
Reception sector in vertical plane, at least, (°).	0–45
Operation control	Local, remote

3 m and, therefore, are placed as stationary variants (in an operating and technical room or a container) (Fig. 2.28b).

The DF coverage area is determined by the range of VHF signal propagation which, in its turn, depends on the flying height of an aircraft and topographic inequalities. The DF coverage area can be reduced due to environmental re-reflections and actions of other interference sources.

The main DF performance specifications are listed in Table 2.2.

Let us consider the design of a DF, its specifications, and operation principle using the example of DF-2000 "*Platan*" [4].

The airdrome and air traffic control tower (ATCT) equipment of DF-2000 "*Platan*" are structurally placed in standard racks. Commercial PCs are applied as digital processors. In addition, the DF composition includes alternate command post (ACP) equipment.

Table 2.3 Main specifications of DF-2000 "*Platan*"

Specification	Value
Frequency band, MHz	100–400
Frequency spacing, kHz	25; 8.33
Type of signal modulation	AM
Number of simultaneously operating channels	2–16
Standard error of direction finding, σ, deg.	1
Range of direction finding, km, at flying height of aircraft	
(150 ± 50) m (300 ± 50) m (1000 ± 50) m (3000 ± 50) m $(10\,000 \pm 50)$ m	$\geq$45 km $\geq$65 km $\geq$120 km $\geq$200 km $\geq$360 km
Signal duration (s)	$\geq$0.5
Coverage in vertical plane (°)	60
Antenna diameter (m)	3.2
Height of antenna mast (m)	5.7

The main specifications of DF-2000 "*Platan*" are listed in Table 2.3.

The DF-2000 is made up of the following main structural and functional components with variable completeness:

an antenna with a test oscillator (TO) used to control the technical condition of the DF-2000;
a main antenna with 16 dipoles spaced circumferentially with the diameter of 3.2 m and with a central dipole, as well as a mast for its installation;
remote indicators;
ATCT equipment which is a unified remote control system (RCE-2000 item). The RCE-2000 equipment provides, if necessary, Ethernet access to the information about the DF condition and parameters;
ACP equipment which duplicates all main functions of RCE-2000 and is placed in ACP rooms;
a box of processing and automatics (BPA).

The DF-2000 can be controlled from the local control panel and remotely (up to 10 km) by means of the RCE-2000 remote control equipment and ACP equipment.

For failure-free operation, the DF-2000 is provided with:

communication lines reservation between the DF airdrome and remote equipment;
processing a signal from the opposite dipole if a dipole of the antenna array fails;
periodic test of direction-finding channels in "OPERATION" mode by means of a signal from the test oscillator;
automatic reservation of direction-finding channels if one of them fails.
The simplified functional diagram of DF-2000 "*Platan*" is shown in Fig. 2.29.

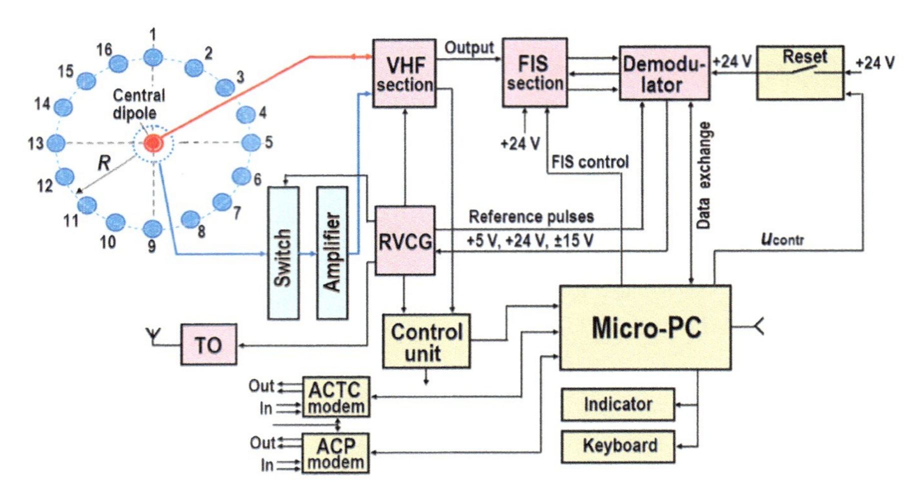

Fig. 2.29 Simplified functional diagram of DF-2000 "*Platan*"

On processing signals, the DF consistently performs the following operations:

multiplexing due to single-band modulation of antenna array signals with the frequency of 5,550 Hz;
transferring a received signal from the carrier frequency to the auxiliary one (5,550 Hz) by means of a demodulator without changing its phase structure;
receiving a measuring signal on the frequency of 43.4 Hz (which corresponds to the frequency of switching the antenna array dipoles) by means of a demodulator;
digitizing the instantaneous values of the measuring signal from each dipole;
calculating the quadrature components of the measuring signal according to cooperative processing of the instantaneous values of signals from 16 dipoles of antenna array;
calculating the initial phase (angle of bearing) and amplitude of the regenerated measuring signal.

The BPA is composed of units for processing, formers of information signals (FIS), high-frequency and powered automatics.

The processing unit includes a micro-PC, modems, a demodulator, a frequency divider—the reference voltage-controlled generator (RVCG).

The built-in micro-PC provides processing the bearing information, control of DF-2000 operation modes, test of its operational integrity, displaying the control results and status.

The modems provide conjugation of the BPA and ATCT/ACP equipment.

The demodulator serves for transferring the spectrum of a direction-finding signal from the frequency of 5,550 to 43.4 Hz (using the phase detection) and for analog-to-digital conversion with the following transfer of processing results to the micro-PC.

The FIS unit includes a distribution amplifier and radio-receiving devices "FIS" according to the number of DF reception channels (from 2 to 16). One unit is meant for eight FISs (channels). The FIS unit provides for switching FIS VHF outputs (via each channel) with connection of either a signal from the simulator of a phase-modulated signal or a direction-finding signal via a frequency channel.

The VHF unit unites the signals of the central dipole and dipoles of the concentric-ring array in a common path; divides the power of the received HF signal according to the number of FIS units; provides two operation modes of the antenna system: "OPERATION" (the central dipole and elements of the antenna system are in the reception mode) and "TEST" (the central dipole is in the mode of radiating the frequency grid in the band of 100.0–150.0 MHz and elements of the antenna array stay in the reception mode).

The powered automatics unit is designed to form and switch the supply voltage; to control the level of secondary supply voltage and main synchronization and control signals; to receive, process and transmit telecontrol signals.

The RCE-2000 ATCT is designed to process and indicate the direction-finding information formed by the DF-2000, as well as for remote control and status monitoring. The RCE-2000 can simultaneously control up to 8 systems (ILS, VOR, DME radio beacons) included in the airdrome ground-based navigation and communication equipment. ATCT can be composed of up to 31 indication modules.

2.4 Summary

Direction-measuring navigation systems such as NDB and ADF have been successfully used in aviation for decades, providing the flight navigation, landing, and surveillance tasks solution. However, the possibilities of their accuracy characteristics improvement are almost depleted and functionality of these systems does not meet the modern requirements to the full extent.

One of the main disadvantages of such systems is the use of direction-measuring method which has the low accuracy of positioning that worsens with its departure from the radio beacon. One more problem of these systems is significant influence of radio-wave propagation's conditions on the accuracy of navigation parameters' determination. That is why such systems serve as a duplication for the more accurate and advanced navigation and surveillance systems.

One expects that in perspective the necessity of using such systems will disappear. However, nowadays and in the mid-term perspective these systems will be in use. This conclusion is based on the low cost of equipment and relative simplicity of these systems, flexibility, and low operational costs as well as on relative high accuracy of these systems. These factors make the systems' usage on remote airfields which are situated in underdeveloped and isolated regions very appealing.

2.5 Further Reading

Additional information about the NDB-ADF system, basic characteristics, principles of construction and functioning of ADF and NDB, the design of Russian production ADF and NDB can be found in [1, 2].

Information on ICAO requirements for NDB parameters and characteristics can be found in [3].

There is a wealth of information on NDB at https://en.wikipedia.org/wiki/Non-directional_beacon.

Additional information about the DF system, basic characteristics, principles of construction and functioning of DF, and the design of Russian production DF can be found in [4].

There is a wealth of information on RDF (DF) at https://en.wikipedia.org/wiki/Radio_direction_finder.

References

1. Skrypnik ON (2014) Radionavigacionnye sistemy vozdushnyh sudov. [Radionavigation systems of aircrafts]. Moscow. INFRA-M (in Russian)
2. Privodnye radiostancii PAR-10S, PARSEK, ARM-150MA, RMP-200. Postroenie i ehkspluataciya. (The homing radio stations PAR-10S, PARSEK, APM-150MA, RMP-200. The construction and operation). Kolomiets VI, Kabanov VA, Tobolov YM. Krasnoyarsk. Siberian branch of the Institute of air navigation. E-learning course (2013) (In Russian)
3. Annex 10 to the Convention on International Civil Aviation. Aeronautical Telecommunications. Volume 1. Radio Navigation Aids. 6-th edition (2006)
4. Avtomaticheskie radiopelengatory ARP-75(80), Platan. Postroenie i ehkspluataciya. [Automatic direction finders ARP-75(80), Platan. The construction and operation]. V.I. Kolomiets, Yu. M. Tobolov. Krasnoyarsk. Siberian branch of the Institute of air navigation. E-learning course (2013) (In Russian)

Chapter 3
Rho-Theta Short-Range Radio-technical Navigation Systems

Rho-theta radio-technical systems of short-range navigation are intended to determine the aircrafts' position at a distance of not more than 400–500 km from the radio beacon. Radio beacons make the content of the ground part—azimuth and range. Toward these radio beacons which are located in the points with known coordinates, aircrafts determine azimuth and slant range. Radio beacons are placed in the air fields' areas and in the points which correspond to the distinctive segments of air routes. Azimuth and range radio beacons operate independently from each other but constructively they can be placed in common instrument room and have combined antennas that have omnidirectional emission in horizontal surface.

Aircraft's coordinate determination is performed by methods of positioning described in Sect. 1.6. Depending on the radio beacons used for this, rho-theta (most frequent), rho-rho (most accurate), or theta-theta (least accurate) methods of positioning are applied.

In world practice, several types of rho-theta short-range navigation systems are used. For civil aviation, ICAO recommends to use VOR/DME system.

VOR/DME system first came into operation in the 1950s, and starting from the 1960s it became the main air radio navigation system. The reason for this was its high accuracy in aircraft's coordinate determination. Nowadays, VOR/DME is still considered to be the main radio navigation system that provides navigation support for existing and for some promising navigation technologies, e.g., RNAV.

Section 3.1 gives common information on existing rho-theta systems of short-range navigation, their main aspects and differences.

In the beginning of Sect. 3.2, VOR system is overviewed: its main characteristics, generalized block diagram and operating principle, mathematical equations, and signals' diagrams are shown that explain the principle of forming the navigation information by the radio beacon and its extraction by onboard equipment. Then, the operating principle and special aspects of antenna system and navigation information formation by the radio beacon DVOR are described. Mathematical equations and signals diagrams are shown that explain the principle of navigation information formation by the DVOR radio beacon.

O. N. Skrypnik, *Radio Navigation Systems for Airports and Airways*,
Springer Aerospace Technology, https://doi.org/10.1007/978-981-13-7201-8_3

Main characteristics of Russian-made azimuthal radio beacons are described, and their engineering aspects are shown. As an example of practical implementation, Russian-made radio beacon DVOR-2000 is overviewed: its construction, main characteristics, functional diagram, and radio beacon's operating principle.

Section 3.3 describes DME system: its main characteristics, generalized block diagram and operating principle, mathematical equations, and signals' diagrams are overviewed that explain principle of signal formation by the radio beacon.

As an example of practical implementation, Russian-made radio beacon DME-2000 is overviewed: its construction, main characteristics, functional diagram, and radio beacon's operating principle.

3.1 Common Characteristics of Rho-Theta Short-Range Radio-technical Navigation System

The rho-theta short-range radio-technical navigation systems (SRRNS) include such systems that provide the aircraft's position determination at a distance of 400–500 km from the radio beacon. This is the result of SRRNS having allocated the UHF and VHF bands in which the operational range is limited by the distance of straight visibility.

The basis of SRRNS is the network of ground-based beacons, and the aircraft determines navigational parameters at—azimuth and/or the distance. In accordance with this azimuth, range and rho-theta radio beacons are distinguished which are installed in the terminal areas and in the points that correspond to the specific areas of air routes. Azimuth and distance-measuring radio beacons operate independently from each other and have omnidirectional antennas emission.

Aircraft's position is determined by LOP method. For this, the information is required about either the distances to two range radio beacons or about azimuths to two azimuth radio beacons or about the azimuth and distance (Fig. 3.1a) to one rho-theta radio beacon (the most common situation). At this, the coordinates of radio beacons must be known in the same coordinate system in which the aircraft's position is being determined as well as the aircraft's true flight height for re-calculating the measured values of slant range into the distance, being measured along the Earth surface.

At the installation on the routes, the cone of ambiguity over the radio beacon is taken into consideration, which depends on the flight height (Fig. 3.1b). In this case, the radio beacon moves from the route for the value of two cones of ambiguity for concrete flight height.

Types of SRRNS are different in the ways of signal formation which is used for azimuth and distance determination as well as in the allocated frequency diapason for the work and system's structure itself. For civil aviation ICAO use, the VHF–UHF diapason for very high-frequency omnidirectional radio range/distance-measuring equipment (VOR/DME) system is recommended. For military aviation

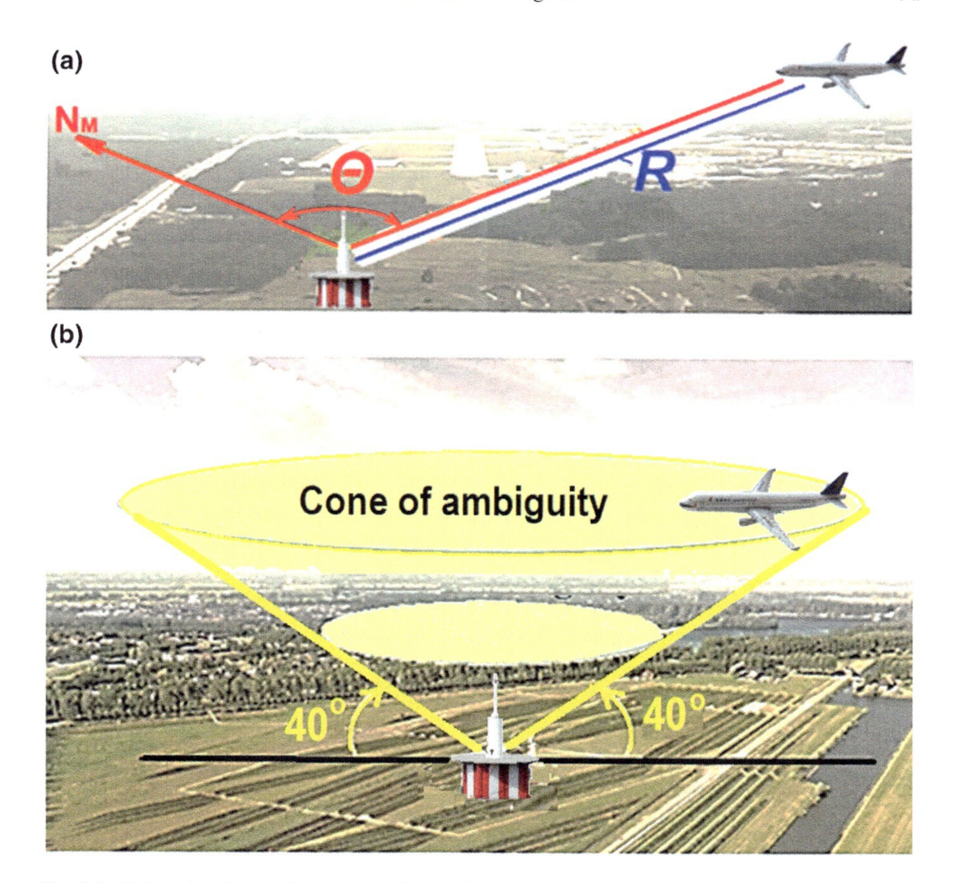

Fig. 3.1 Principle of aircraft's position determination

use, the UHF diapason for TACtical Air Navigation system, USA (TACAN) and Radiotehnicheskaja Sistema Blizhnej Navigacii, Russia (RSBN) is allocated.

VOR/DME system is based on the phase method of azimuth determination and temporal (pulse) method of distance determination. Ground part of the system is formed by azimuth radio beacons VOR (DVOR), and range radio beacons DME that can be used both simultaneously and individually, forming azimuth or range SRRNS.

Azimuth radio beacon operates in VHF diapason (108–118 MHz) while range one operates in UHF (960–1,215 MHz).

On board of an aircraft stand-alone devices are used for azimuth and distance determination while working with VOR and DME radio beacons.

RSBN system uses temporal (pulse) method for determining azimuth as well as the distance. Azimuth and range RSBN radio beacons are united and constructively make a unified complex of ground-based equipment—rho-theta radio beacon.

In RSBN system, the frequency diapason 873.6–1,000.5 MHz is allocated for the radio beacons' signals emission. In modernized variant of RSBN (Tropa-SMD), the diapason of frequency emission is in compliance with international standards and makes 962.0–1,000.5 MHz. This system can be equipped with the second range channel (*DME-Tropa* apparatus) that uses the frequency diapason 960–1,215 MHz.

A distinctive feature of RSBN is a possibility of ground-based air situation surveillance, coordinates' determination, and aircraft's navigation identification for the benefits of air traffic management.

On board of an aircraft, the common device is used for azimuth and distance determination while working with RSBN radio beacons. In the same device, the possibility of signals' receipt from landing beacons that are operating on rho-theta radio beacons frequencies is covered. For this reason, RSBN can be considered as a complex system of short-range navigation and landing. At having the specific calculators as a part of onboard apparatus, the aircraft's position determination task, as well as the correction of calculated (dead reckoning) coordinates can be solved. Some types of onboard equipment RSBN enable to solve the en route air navigation task (aircraft's encounter, group gathering, piloting within a group) by calculating relative distances and azimuths.

TACAN system is close to RSBN system by its operation principle, frequencies diapason, and purpose (except for the landing function), but the azimuth channel uses the phase method of measuring. In some variants of TACAN, the ground-based observation of the air situation is possible. A rho-theta radio beacon and onboard equipment that measures azimuth and the distance of aircraft are a part of the system. The system VORTAC is also well-known, and it combines VOR radio beacon and the range radio beacon of TACAN system.

3.2 VOR System

3.2.1 Main Characteristics and the Destination of System

VOR system is created for determining the aircraft's azimuth toward the ground-based radio beacon—an angle between the direction of North magnetic meridian passing through the radio beacon's antenna system and the direction to the aircraft, counted clockwise. Azimuth's determination makes it possible to use VOR for the following tasks:

marking the beginning, the end, and centerline of airways or sections of airways;
as an auxiliary aid at airfields using standard approach procedures;
as a reference point for aircraft;
as a source of navigation information (lines of position) for aircraft position fixing.

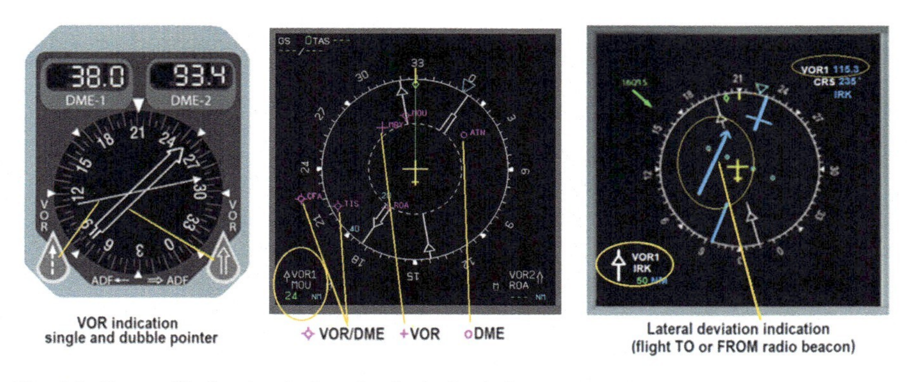

Fig. 3.2 Types of indicating devices for displaying information from VOR

By measuring azimuths to two radio beacons, which coordinates are known, the aircraft's position can be determined using the theta-theta method. It is important to note that the accuracy of such position determination will be relatively low.

VOR radio beacon forms and emits signals with horizontal polarization through non-directional antenna system, the class of emission is A9W.

The VOR radio beacon emits 360 radio beams (lines of position or radials) that are divided by 1° intervals which enables pilot to know on which radial he is flying TO or FROM the VOR radio beacon.

In order to identify, VOR radio beacon transmits an identification signal on the same VHF carrier which is used to provide the navigational function. In order to transmit the identification signals, AM carrier by two- or three-letter Morse code is used. This signal is transmitted with the speed of approximately seven words per minute and repeats every 30 s. The frequency of modulation's tone signal is 1,020 ± 50 Hz. Identification signal of radio beacon can also be transmitted via voice message.

Along with its main function execution, VOR radio beacon can provide the communication channel "earth-air" on the same VHF carrier which is used for navigational function execution. Radiotelephony communication does not interrupt the execution of a main VOR navigation function. Identification signals are not concealed at signals' emission.

Onboard equipment of VOR is the receiver-indicator which, besides of the azimuth indication, allows to guide aircraft in FROM or TO the radio beacon modes following the set azimuth, as well as to determine the side and value of lateral deviation from the line of set path. On the different aircraft's types, different types of indication devices can be used (Fig. 3.2).

On practice, the following types of VOR radio beacons are used:

CVOR—conventional VOR is used to define airways and for en route navigation;
DVOR—a Doppler VOR (this overcomes re-reflected signals errors);

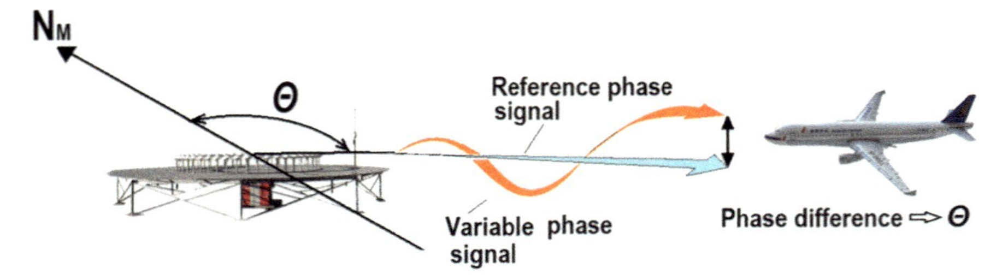

Fig. 3.3 VOR (DVOR) system operating principle

TVOR—terminal VOR which has only low output power (50 W) and is used in the terminal area of airports and covers a relatively small earth's surface protected from interference by other stations on the same frequency band;
VOT—this is found at certain airfields and broadcasts a fixed omnidirectional signal for a 360° test radial. This is not for navigation use but is used to test an aircraft's equipment accuracy before flight. More than ±4° indicates that equipment needs servicing.

Allocated frequency diapason for VOR operation is divided into two parts: from 108 to 112 MHz (contains 40 channels), and from 112 to 117.95 MHz (contains 120 channels with step 50 kHz).

Frequency diapason 108–112 MHz is primarily an ILS band but ICAO has allowed it to be shared with short-range VORs and Terminal VORs: 108.0, 108.05, 108.20, 108.25, 108.40, 108.45, ...,111.85 MHz (even decimals and even decimals plus 0.05 MHz).

En route radio beacons are used in the frequency band 112–117.95 MHz.

Phase method of azimuth measurement lies in the basis of VOR operating principle. Thereat, the information carrier can be either the envelope phase of AM fluctuations or the phase of high-frequency carrier.

In systems of the first type, dependence of AM fluctuation's envelope phase from the azimuth is created by rotating the low-directivity polar diagram of a radio beacon's antenna system. Cardioid-type polar diagram is normally used.

In systems of the second type, phase dependence of VHF carrier from azimuth is created by rotating non-directional antenna in a circumferential direction of specified radius. Thereat, Doppler effect is occurred and used for azimuth determination, which leads such systems to be called Doppler systems.

In both types of system, radio beacon emits the signal of reference phase as well as the signal, which phase of modulating fluctuation or frequency carrier depends on the azimuth (azimuth signal) (Fig. 3.3). Reference phase and azimuth's signals phases are the same and equal to zero when either the cardioid minimum or the position of rotating antenna are pointed to the magnetic north. Onboard equipment processes these signals and measures the difference between azimuth and reference signals' phases, which is proportional to azimuth.

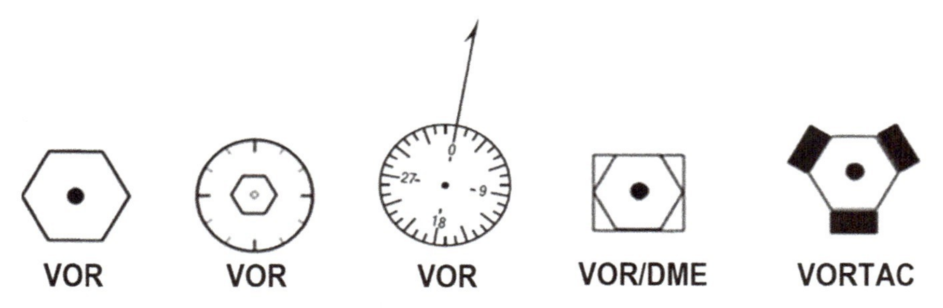

Fig. 3.4 Radio beacons' conventional signs

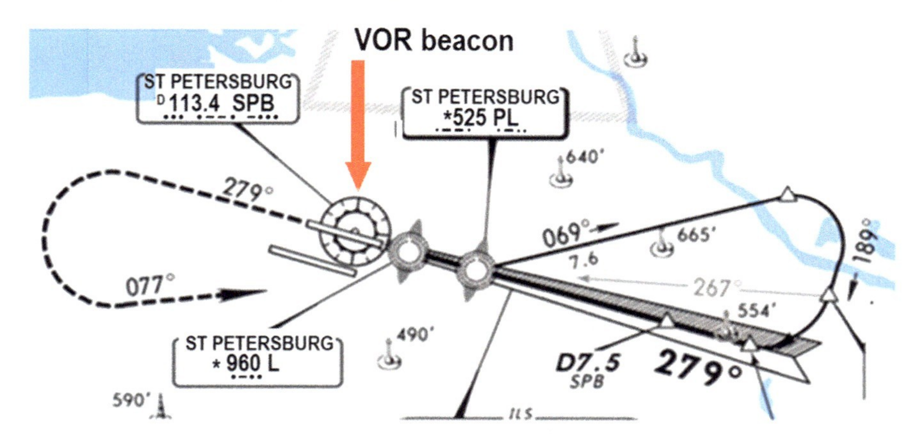

Fig. 3.5 VOR image on the navigation map

VOR radio beacons' conventional signs in combination with DME and Tacan radio beacons, which are used on the aeronautical charts, are shown in Fig. 3.4.

The "box" is put on the navigational map (Fig. 3.5)—a rectangle containing data that identifies VOR radio beacon. This data include:

VOR's (ST PETERSBURG) title that matches the town or any other ground point where it is located;
frequency (113.4 MHz);
call sign (SPB) that can be duplicated by Morse code or voice message;
Letter D before the frequency value indicates that VOR radio beacon is combined with DME.

On some maps, geodesic coordinates of radio beacon can also be specified (on Fig. 3.5, this information is missing).

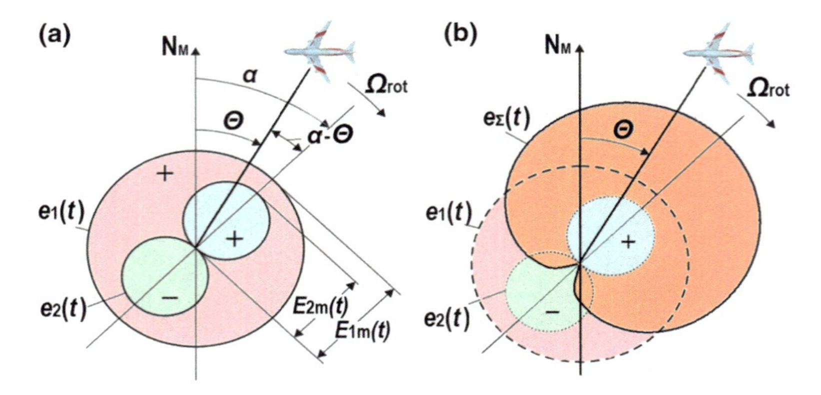

Fig. 3.6 Angular patterns of VOR radio beacons antennas

3.2.2 VOR System's Operating Principle

Operating principle of ***standard VOR*** is based on AM signal formation, where the information about azimuth is enclosed in modulation envelope phase. Such type of signal is created in space using antenna system that consists of omnidirectional antenna A_1 with polar diagram $F_1(\alpha) = 1$ and directional antenna A_2 with polar diagram $F_2(\alpha) = \cos\alpha$ which rotates in horizontal plane with the frequency $F_{\text{rot}} =$ 30 Hz (Fig. 3.6a).

In any azimuth direction, Θ value of A_2 antenna's angular pattern is characterized by the $F_2(\alpha - \Theta) = \cos(\alpha - \Theta)$ value.

Antenna A_1 creates a field:

$$e_1(t) = E_{m1}\cos\,\omega_0 t, \tag{3.1}$$

where E_{m1} amplitude and $\omega_0 t$ phase do not depend on the azimuth.

A_2 antenna creates a field in any azimuth direction Θ as:

$$e_2(t) = E_{m2}\cos(\alpha - \Theta)\cos\omega_0 t, \tag{3.2}$$

with $E_{m2}\cos(\alpha - \Theta)$ amplitude that depends on azimuth.

Normally, the condition $E_{m1} > E_{m2}$ is fulfilled for the VOR radio beacons.

Antennas have a common phase center. The resultant radiation field of radio beacon in space is created by composing the fields (3.1) and (3.2) produced by antennas, which aerial system's polar diagram has cardioid form (Fig. 3.6b) as:

$$e_\Sigma(t) = e_1(t) + e_2(t) = E_{m1}\left[1 + \frac{E_{m2}\cos(\alpha - \Theta)}{E_{m1}}\right]\cos\omega_0 t. \tag{3.3}$$

A_2 antenna's polar diagram revolves in horizontal plane with the angle speed Ω_{rot} that matches the frequency $F_{\text{rot}} = 30$ Hz.

The position of A_2 antenna's polar diagram in horizontal plane (its maxima position) makes the temporal function $\alpha = \Omega_{\text{rot}} t$. Antenna's rotation causes recurrent resultant radiation field's change (3.3).

Let us define the amplitude ratio $E_{m2}/E_{m1} = m_s$ and by putting the m_s and $\alpha = \Omega_{\text{rot}}\, t$ values to the (3.3), we will get the following equation:

$$e_{\Sigma}(t) = E_{m1}[1 + m_s \cos(\Omega_{\text{rot}} t - \Theta)] \cos \omega_0 t. \tag{3.4}$$

As a result of radiation fields' composition that are formed by A_1 and A_2 antennas, the radiation field with the spatial depth AM m_s, modulation frequency $\Omega_{\text{rot}} = 2\pi F_{\text{rot}}$, and modulation envelope phase which depends on the Θ azimuth is formed in space.

The signal which is being received by onboard receiver can be presented using the following equation:

$$u_{\text{rec}}(t) = U_m[1 + m_s \cos(\Omega_{\text{rot}} t - \Theta)] \cos \omega_0 t. \tag{3.5}$$

After amplification and detection, low-frequency voltage can be extracted (signal of alternating phase, azimuth signal) from (3.5) as:

$$u_s(t) = m_s U_m \cos(\Omega_{\text{rot}} t - \Theta), \tag{3.6}$$

which phase $\varphi_s(\text{t}) = \Omega_{\text{rot}} t - \Theta$ contains the information about the plane's azimuth Θ.

For extracting this information, it is required to have reference signal in onboard receiver, carrying the information about the immediate position of a rotary antenna's polar diagram. This information must be put in the phase of reference signal:

$$u_{\text{ref}}(t) = U_{m\text{ref}} \cos \Omega_{\text{rot}} t, \tag{3.7}$$

with the current phase value $\varphi_{\text{ref}}(t) = \Omega_{\text{rot}} t$, matching angular position of antenna's A_2 polar diagram in the moment of time t.

Given such reference signal on board of an aircraft, the azimuth can be determined as the difference between reference (3.7) and azimuth (3.6) signal phases $\Theta = \varphi_{\text{ref}}(t) - \varphi_s(\text{t})$.

Thus, for extracting the information about azimuth, the reference signal is required, whereby the same for all aircrafts. The problem arises of how to organize the channel of reference signal transmission and not to expend the frequency resource of a system.

For this, the reference signal is being transmitted on the same carrier frequency ω_0 as the azimuth one. The division of azimuth and reference signals by the channels occurs on the receiving side by using the frequency selection method of mixed signal

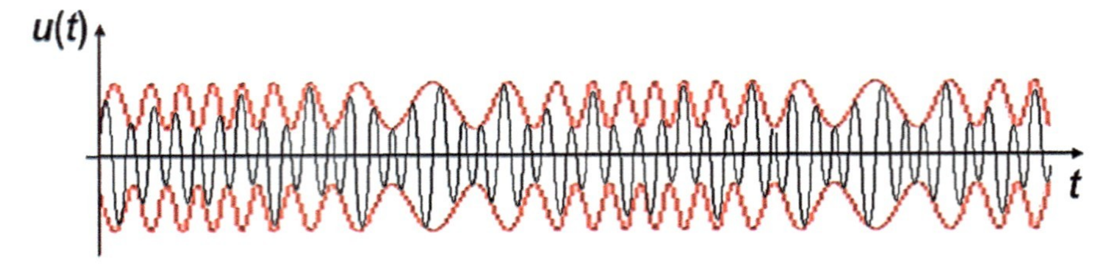

Fig. 3.7 Type of signal with dual-amplitude-frequency modulation

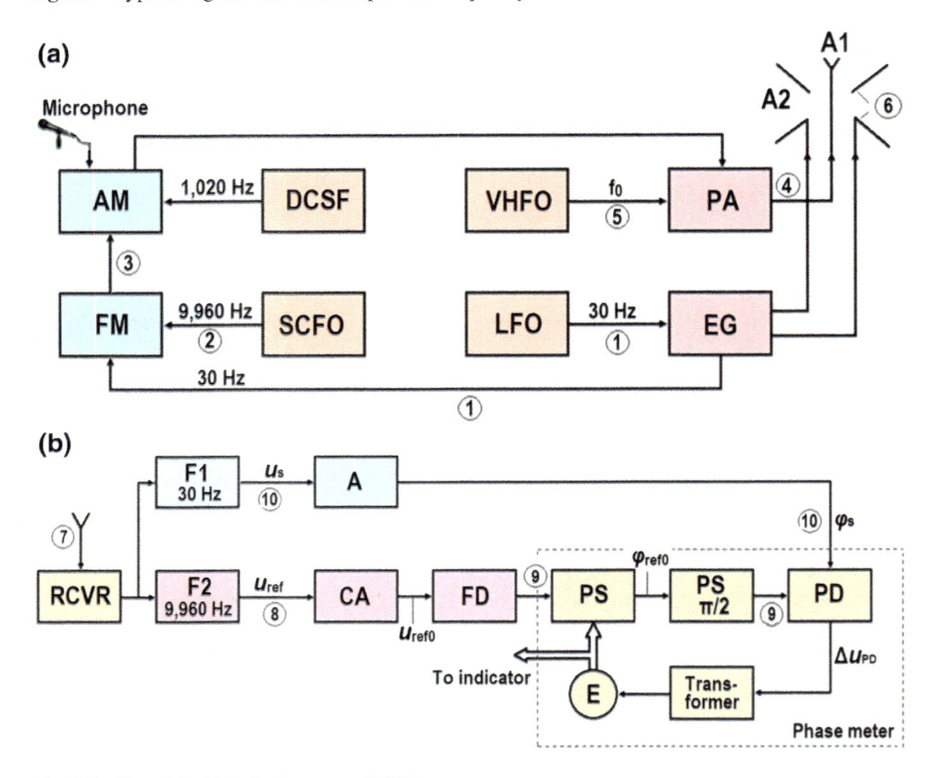

Fig. 3.8 Simplified block diagram of VOR system

rectified by amplitude. Such opportunity arises at using the dual-amplitude-frequency modulation for reference signal transmission (Fig. 3.7).

Beside the reference and azimuth signals, the radio beacon emits call signals and voice signals (identification or traffic controller's commands) transmission possible through its transmitter.

Let us take a look at the signal formation by the onboard equipment and the performance of onboard equipment on the example of simplified block diagram of VOR system (Fig. 3.8a—radio beacon, Fig. 3.8b—receiver).

Antenna system of a radio beacon consists of static omnidirectional antenna A_1 and antenna A_2 that creates rotating in horizontal plane polar diagram.

In early radio beacons, the rotation of A_2 antenna's polar diagram was achieved due to the mechanical rotation of the antenna itself with the help of electric motor. Later, methods of electronic rotation of static antenna's polar diagram were developed. The effect of polar diagram's rotation is achieved by using the electronic goniometer (EG)—device that consists of two mutually perpendicular directional antennas with polar diagrams in a shape of "8", powered by balance-modulated oscillations with 90° phase difference. The frequency of polar diagram's rotation is defined by the frequency of balance-modulated oscillations and makes 30 Hz.

Loop antennas have polar diagrams in a shape of "8", as well as the vertical omnidirectional dipoles that are switched on in a reverse-phase manner.

Balance-modulated signal, transmitted by A_2 antenna is a signal of "variable phase" that carries information about azimuth of an aircraft toward the radio beacon installation point as:

$$e_2(t) = E_{m2}\cos(\Omega_{\text{rot}}t - \Theta)\cos\omega_0 t.$$

In order to form the signal with double amplitude-frequency modulation, the fluctuations are chosen which frequency is much greater than the rotation frequency of antenna A_2 polar diagram, but is less than the carrier wave frequencies and these fluctuations are used as the secondary ones. Secondary fluctuations are called the subcarrier, for which the following condition must be fulfilled $\Omega_{\text{rot}} \ll \Omega_{\text{sc}} \ll \omega_0$ where Ω_{sc} is the frequency of subcarrier waves. Such fluctuations are produced by a generator of subcarrier waves (subcarrier frequency oscillator—SCFO). For VOR system, the frequency of a subcarrier is $F_{\text{sc}} = 9{,}960$ Hz.

In frequency modulator (FM), the subcarrier's frequency modulation is performed by reference waves with frequency $F_{\text{ref}} = F_{\text{rot}} = 30$ Hz and frequency deviation $\Delta F_{\text{sc}} = 480$ Hz. In the power amplifier (PA), this signal modulates carrier's fluctuations, and the resultant VHF signal with double amplitude-frequency modulation, carrying the information about reference phase, comes to A_1 antenna. A_1 antenna creates a radio emission field with electric intensity:

$$e_1(t) = E_{m1}\left[1 + m_{\text{sc}}\cos\left(\Omega_{\text{sc}}\,t + \frac{\Delta\Omega_{\text{sc}}}{\Omega_{\text{rot}}}\cos\Omega_{\text{rot}}\,t\right)\right]\cos\omega_0\,t, \tag{3.8}$$

where m_{sc} is the amplitude modulation coefficient; $\Delta\Omega_{\text{sc}}/\Omega_{\text{rot}}$ is the frequency modulation coefficient; $\Delta\Omega_{\text{sc}}$—deviation of subcarrier's frequency.

As a result of electromagnetic fields in space composition which are transmitted by A_1 and A_2 antennas, the resultant polar diagram is formed in the cardioid shape that is rotating with 30 Hz frequency.

The sources of modulating fluctuations are the low-frequency (30 Hz) oscillator (LFO) and the device for call sign formation (DCSF). Amplitude modulator (AM) can also be used for transmitting the voice signals from the ATM system controller onto the board of an aircraft.

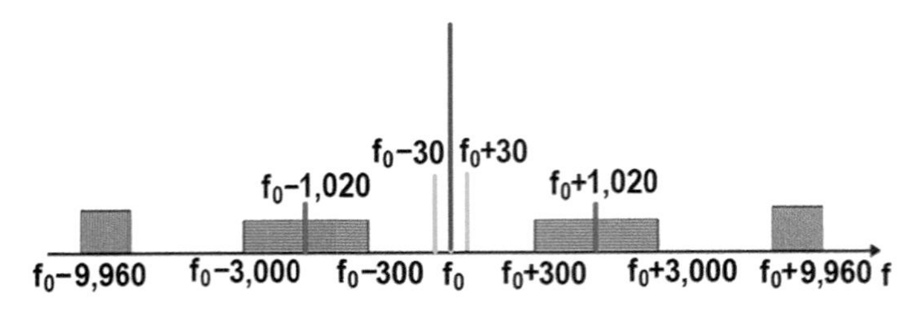

Fig. 3.9 VOR radio beacon's complete signal spectrum

Resultant field impacts the antenna of aircraft's receiver VOR. At the output of antenna, the resultant fluctuation of the following type is formed as:

$$u_{\Sigma}(t) = U_{\Sigma m}\left[1 + m_s \cos(\Omega_{\text{rot}} t - \Theta) + m_{\text{sc}} \cos\left(\Omega_{\text{sc}} t + \frac{\Delta\Omega_{\text{sc}}}{\Omega_{\text{rot}}} \cos \Omega_{\text{rot}} t\right)\right] \cos \omega_0 t. \tag{3.9}$$

Line spectrum of the VOR radio beacon's complete signal is illustrated in Fig. 3.9.

Composite signal spectrum (3.9) contains the following components:

frequency carrier waves f_0;
two side frequencies $f_0 \pm 30$ Hz;
two side frequency bandwidths $f_0 \pm (3{,}000 \pm 300)$ Hz—carrier that is amplitude-modulated by voice signals, transmitted via radio beacon;
two side frequencies $f_0 \pm 1{,}020$ Hz—carrier that is amplitude-modulated by Morse code signals;
two side frequency bandwidths $f_0 \pm (9{,}960 \pm 480)$ Hz—carrier waves that are amplitude-modulated by 9,960 Hz subcarrier which is in its turn modulated by frequency.

After resultant signal's conversion in the receiver, its amplification and the amplitude detection, the signal envelope that contains the azimuth and reference signals of the type which is described below is being extracted as:

$$u_{\text{env}}(t) = U_{m2} \cos(\Omega_{\text{rot}} t - \Theta) + U_{m1} \cos\left(\Omega_{\text{sc}} t + \frac{\Delta\Omega_{\text{sc}}}{\Omega_{\text{rot}}} \cos \Omega_{\text{rot}} t\right), \tag{3.10}$$

where U_{m1} and U_{m2} are the amplitudes of complete signal's components.

Azimuth and reference signals can be divided by frequency selection. In order to do that, the signal is transmitted to two filters F_1 and F_2 at the receiver's output.

In the F_1 filter that is configured to frequency Ω_{rot} ($F_{\text{rot}} = 30$ Hz), azimuth signal or signal of alternating phase $u_s(t) = U_{m2} \cos(\Omega_{\text{rot}} t - \Theta)$ is detected and then amplified in the amplifier (A).

In F_2 filter that is configured to subcarrier's frequency Ω_{sc} ($F_{\text{sc}} = 9{,}960$ Hz), frequency-modulated signal of subcarrier is detected. After symmetrical limiting in

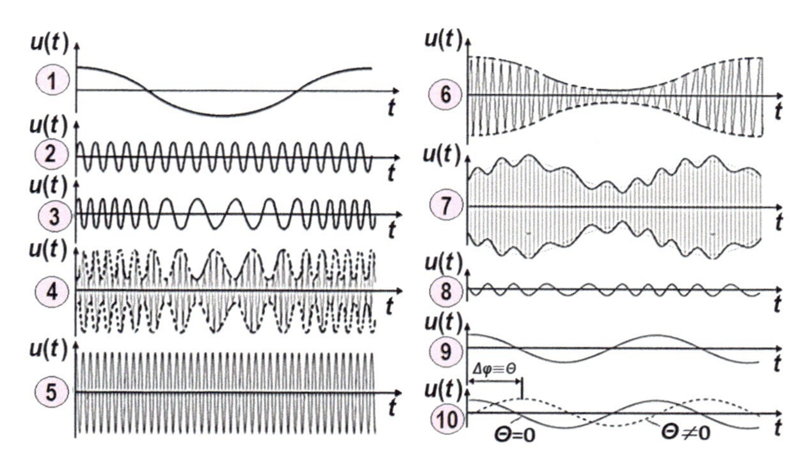

Fig. 3.10 Diagrams of VOR system signals

clipper amplifier (CA), the reference signal $u_{\mathrm{ref}}(t) = \mathrm{U}_{m\mathrm{ref}} \cos \Omega_{\mathrm{rot}}t$ is being detected in frequency detector (FD).

Azimuth and reference signals are transmitted on phase meter (phase detector, PD), which output signal contains information about azimuth. This signal is then conversed into the type which is required for transmitting onto a device of azimuth indication.

In Fig. 3.10, the diagram of signals on output of main functional radio beacon units and aircraft receiver are illustrated, which explain the subject of navigational information forming and processing in VOR system.

On Fig. 3.10 are shown: 1—signal of reference phase (30 Hz frequency); 2—signal of subcarrier (9,960 Hz frequency); 3—frequency-modulated signal of subcarrier; 4—composite signal, which is emitted by central omnidirectional antenna A_1; 5—signal, which is emitted by rotating antenna A_2; 6—signal, which is received on board the aircraft from the rotating antenna (initial phase of this signal's envelope depends on azimuth); 7—composite signal at the input of receiver. It is a VHF signal, which is amplitude modulated by subcarrier's signal, which in its turn is frequency modulated by the reference signal. In this signal, the amplitude modulation is also present by signal of variable phase (azimuth signal) (30 Hz frequency); 8—voltage at the output of F_2 filter—signal of subcarrier, which is frequency modulated by reference signal; 9— signal at the output of PD—reference signal; 10—signal at the output of F_1 filter—signal of variable phase.

In VOR system, the possibility of aircraft's flight on the set path (azimuth) Θ_{def} is foreseen. For this purpose, the schema that provides the reference signal's phase movement for the value that matches the set path and the second phase detector PD_2 is introduced in the receiver.

Thereat, the reference signal's phase u_{ref} is additionally moved for the Θ_{def} value and makes $\Omega_{\mathrm{rot}}t - \Theta_{\mathrm{def}}$.

The azimuth signal with the phase that depends on the azimuth $\Omega_{\text{rot}}t - \Theta$ comes to the second input of phase detector.

The difference in azimuth and reference signals' phases at the FD_2 entries:

$$\Delta\varphi = \Omega_{\text{rot}}t - \Theta_{\text{def}} - (\Omega_{\text{rot}}t - \Theta) = \Theta - \Theta_{\text{def}} = \Delta\Theta.$$

At the output of detector, the following voltage is formed:

$$u_{\text{PD}} = U_m \sin\Delta\varphi = U_m \sin\Delta\Theta. \tag{3.11}$$

When $u_{\text{PD}} = 0$, then $\Delta\Theta = 0$ and $\Theta = \Theta_{\text{def}}$, i.e., aircraft's azimuth matches the specified direction. This task is solved at the aircraft's flight TO or FROM the VOR radio beacon.

In order to perform the flight indication TO or FROM the radio beacon, the third phase detector PD_3 is implemented into the scheme of a receiver, on which the azimuth signal $\text{u}_{\text{s}}(t) = \text{U}_{m2}\cos(\Omega_{\text{rot}}t - \Theta)$ with phase $\varphi_{\text{s}} = \Omega_{\text{rot}}t - \Theta$ and the reference signal $u_{\text{ref}}(t) = U_{m\text{ref}}\cos(\Omega_{\text{rot}}t - \Theta_{\text{def}} + 90°)$ with phase $\varphi_{\text{ref}} = \Omega_{\text{rot}}t - \Theta_{\text{def}} + 90°$ are transmitted.

The difference between these signals' phases is $\Delta\varphi = \Theta - \Theta_{\text{def}} + 90°$.

While flying to the radio beacon and having $\Theta = \Theta_{\text{def}}$ at the output of PD_3, according to (3.11), we will have $u_{\text{PD}} = +U_m$. Positive voltage causes activation of the indicating panel "TO".

After passing the radio beacon, current aircraft's azimuth changes for 180°, and $\Theta = \Theta_{\text{def}} \pm 180°$. Azimuth change causes the change of voltage polarity at the output PD_3 $u_{\text{PD}} = -U_m$, therewith the indicating panel "TO" turns off and the panel "FROM" turns on.

Error rate of azimuth measurement in VOR system considerably depends on the terrain configuration (presence of reflections from the relief roughness, buildings, and constructions in the terminal area) which is the main disadvantage of this system. Moderate decrease of reflections' impact provides the appliance of emitted signals' horizontal polarization.

3.2.3 DVOR System's Operating Principle

Doppler VOR (DVOR) system is a modernization of VOR radio beacon and due to the usage of Doppler effect and an antenna with a large base (radius of rotating), it can provide significantly higher accuracy of azimuth measuring. DVOR is normally used in the areas with difficult terrain relief.

In order to explain DVOR operating principle, let us consider the antenna system that consists of the central and lateral antennas (Fig. 3.11). Central antenna A_c is motionless, while the lateral antenna A_s rotates toward the central one around the

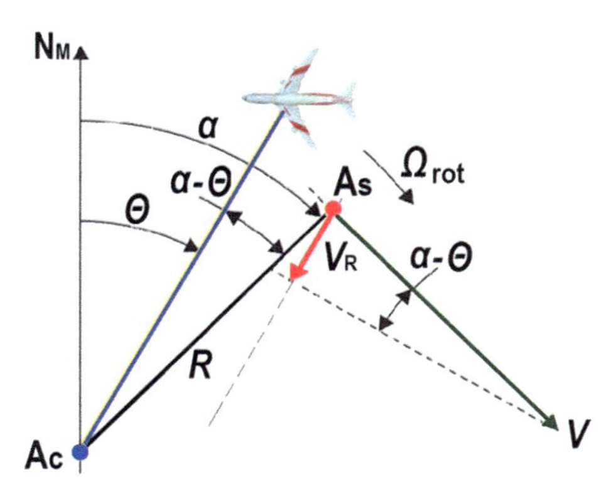

Fig. 3.11 Doppler effect usage in DVOR

radius R with the rotation speed Ω_{rot}. Current position of lateral antenna is characterized by angle α, aircraft's azimuth by angle Θ.

Antenna's linear rotation speed is $V = \Omega_{\text{rot}} R$. Then, the radial component of speed in the direction of Θ is:

$$V_{\text{r}} = V \sin(\alpha - \Theta). \tag{3.12}$$

Current angular position of lateral antenna is $\alpha = \Omega_{\text{rot}} t$. When $t = 0$, antenna is situated on the base line of counting (it takes position in the direction of magnetic North) and $\alpha = 0$. Taking into account (3.12), radial speed is:

$$V_{\text{r}}(t) = \Omega_{\text{rot}} R \sin(\Omega_{\text{rot}} t - \Theta). \tag{3.13}$$

When it comes to the signal reception from the rotating lateral antenna A_{s}, a Doppler frequency shift arises in the reception point $F_{\text{D}} = V_{\text{r}}/\lambda$ (λ—wave length of receivable fluctuations).

Considering expression (3.13), we will get the following equation:

$$F_{\text{D}}(t) = \frac{\Omega_{\text{rot}} R \sin(\Omega_{\text{rot}} t - \Theta)}{\lambda}.$$

Let us point out that $\frac{\Omega_{\text{rot}} R}{\lambda} = F_{\text{Dmax}}$, then:

$$F_{\text{D}}(t) = F_{\text{Dmax}} \sin(\Omega_{\text{rot}} t - \Theta). \tag{3.14}$$

It is clear from the (3.14) that the Doppler frequency shift $F_{\text{D}}(t)$, obtained as a result of lateral antenna's rotation, contains information about aircraft's azimuth Θ.

Obtaining a signal from rotating antenna to the aircraft, information (azimuth) signal of the following type can be extracted as:

$$u_s(t) = U_{s\,\max} \cos(\Omega_{\text{rot}} t - \Theta),$$

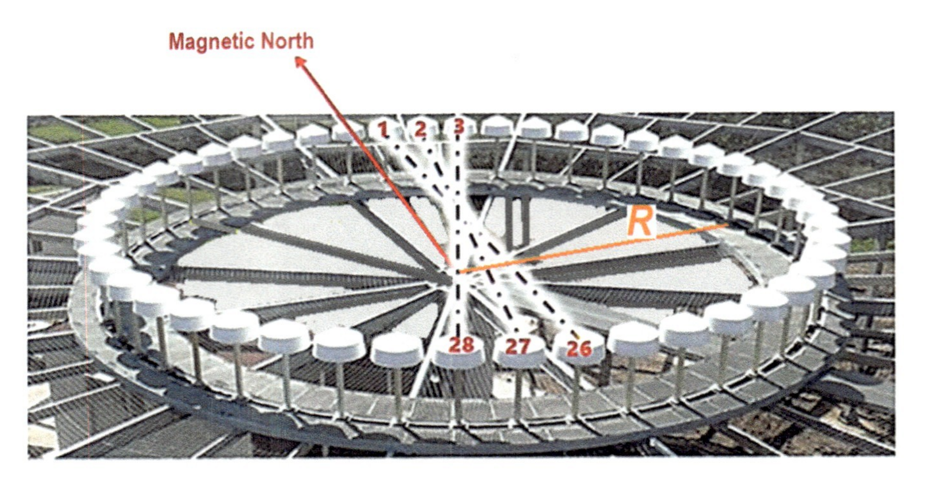

Fig. 3.12 DVOR radio beacon's antenna system

which phase depends on azimuth.

Reference signal

$$u_{\text{ref}}(t) = U_{\text{ref max}} \cos \Omega_{\text{rot}} t,$$

which phase does not depend on azimuth is required for the information extraction about azimuth.

This signal (reference phase signal) is transmitted via the central antenna A_c. When comparing the phases of reference and azimuth signals in onboard equipment azimuth of an aircraft is determined.

Main advantage of Doppler radio beacons, comparing to the standard VOR radio beacons is significant reduction of local items influence on the accuracy of azimuth determination. For effective suppression of multipath signals, rotation radius of R antenna must be relatively big ($2R/\lambda = 4{-}6$), while the speed of direction finding must stay high. This requires antenna to rotate at a high-frequency rate which is difficult to achieve when it is mechanical rotating. That is why in modern DVOR systems motionless antenna arrays that consist of a lot of emitters (dipoles) situated circumferentially are used instead of rotating antennas. Rotating effect is achieved by sequential commutation of emitters. Thereat, the format and spectral signals' structure of DVOR radio beacons must match the VOR radio beacons in order to have a possibility of their reception and processing with the help of single-type onboard equipment without its modification or replacement.

DVOR radio beacon antenna's system (Fig. 3.12) consists of a big number, e.g., 50, of emitters, that are located equally in a circular motion of radius R (approx. 6–7.5 m).

Opposite emitters (1 and 26, 2 and 27, etc.) power currents with frequencies $f_k = f_0 + F_{\text{sc}}$ and $f_{k+25} = f_0 - F_{\text{sc}}$ accordingly, where f_0 is a wave carrier, F_{sc}–subcarrier frequency of VOR system which is equal to 9,960 Hz, $k = 1, \ldots, 25$—the number of dipole. Sequential connection of emitters pairs to the transponder of radio beacon imitates its rotating in a circular motion with rotating frequency $F_{\text{rot}} = 30$ Hz.

Simultaneous signals' transmission by direct opposite emitters allows to obtain a signal in space with two modulation frequency side bands ±9,960 Hz, which is equivalent to VOR signal.

Transmitters are mounted over the screen of 30 m in diameter which is made for decreasing the influence of earth terrain on the properties of antenna system's polar diagram.

Signals, received on board of an aircraft have frequencies $f_k = f_0 + F_{\text{sc}} \pm F_{\text{D}}$ and $f_{k+25} = f_0 - F_{\text{sc}} \pm F_{\text{D}}$ due to the Doppler shift, i.e., signals are modulated by frequency with frequency change. Thereat, Doppler's frequency shift sign depends on whether connection of this emitters leads to the increase or decrease in terms of the distance to aircraft.

Reference signal is transmitted through the central antenna, which features AM fluctuations:

$$e_{\text{c}}(t) = E_{\text{c}\max}(1 + m_{\text{ref}} \cos \Omega_{\text{rot}} t) \cos \omega_0 t.$$

As a result of central antenna and lateral emitters fields' composition, the signal is formed in the reception point as:

$$e_{\Sigma}(t) = E_{c\max}[1 + m_{\text{ref}} \cos \Omega_{\text{rot}} t + m_{\text{sc}} \cos(\Omega_{\text{sc}} t + m_{\text{FM}} \cos(\Omega_{\text{rot}} t - \Theta))] \cos \omega_0 t, \tag{3.15}$$

where $m_{\text{FM}} = 2\pi R/\lambda$ is the index of frequency modulation.

It is clear from the (3.15) that this signal is identical to the standard VOR signal by its structure. The main difference of DVOR radio beacon's signal processing in onboard equipment is that the azimuth signal (signal of variable phase) is transmitted via FM channel and is produced by F_2 filter (Fig. 3.8), while the reference signal is transmitted via AM channel and is produced by F_1 filter in the onboard equipment.

Diagrams of signals which explain DVOR system operating principle are illustrated in Fig. 3.13.

In Fig. 3.13 marked: 1—signal of variable phase (azimuth signal) with frequency 30 Hz; 2—signal of subcarrier's frequency modulator 9,960 Hz; 3—signal at the output of antenna array (modulated by amplitude with 9,960 Hz frequency); 4—signal on board of the aircraft (frequency of carrier and modulation is changed due to Doppler effect); 5—signal of reference phase frequency 30 Hz (emitted by central antenna).

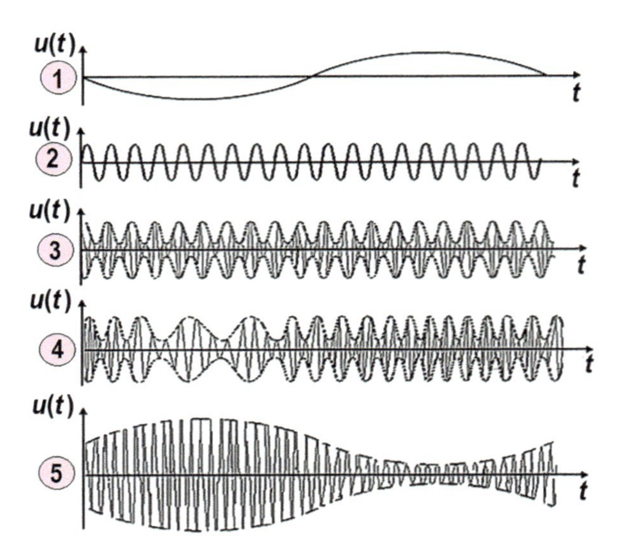

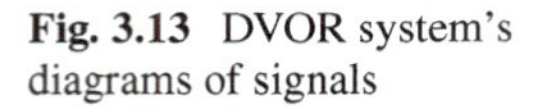
Fig. 3.13 DVOR system's diagrams of signals

3.2.4 Construction and Main Technical Characteristics of VOR (DVOR) Radio Beacons

Main outside and visible difference of VOR and DVOR radio beacons is the construction of antenna systems.

VOR radio beacons have relatively small antenna's system dimensions (Fig. 3.14a, RMA-95 radio beacon, Russia), while DVOR antenna system (Fig. 3.14b, DVOR-2000 radio beacon, Russia) features an overall and complicated construction.

In Table 3.1, main characteristics and parameters of azimuth radio beacons made in Russia are listed.

Let us take a look at the construction, characteristics, and operating principle of azimuth radio beacons on the example of DVOR-2000 (Russia) [3].

DVOR-2000 apparatus has the following main features:

antenna system of a radio beacon contains 48 emitters that are located circumferentially of 13.5 m diameter which together with two side frequency emission $F_0 \pm$ 9,600 Hz by opposite emitters provides minimal deviations of subcarrier;
control system provides real-time parameters monitoring of radio beacon's main functional modules and the quality of emitted signal in near area at its receipt by controlling antenna of a near area. Signals are then processed and analyzed by monitor's processor using the discrete Fourier transformation. Control system estimates the tendencies of radio beacon parameters' change as well;
modulation signals generation, VHF signals' phase and amplitude control is achieved with the help of microprocessors, which improves maintenance-technical characteristics of apparatus;

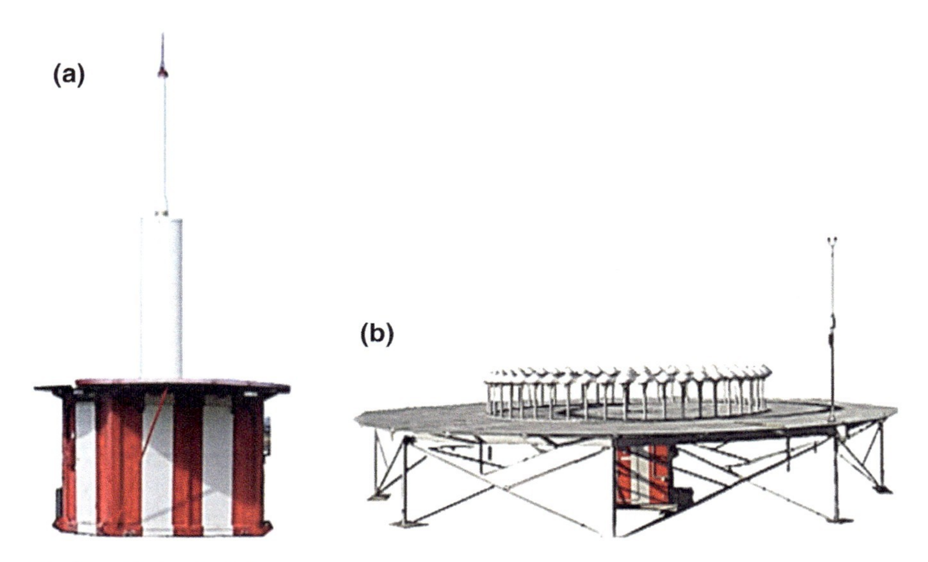

Fig. 3.14 VOR and DVOR visual appearance

module construction of apparatus as well as a wide usage of DVOR-2000 series submodules are realized in the radio beacon. All submodules of transmitters and monitors are mounted in common stand (box).

Radio beacon consists of airfield apparatus:

operating room, where the apparatus of radio beacon's signal forming, control, and monitor is placed, as well as the radio-receiver P-872B, designed for signal receipt from controlling antennas;
antenna system that consists of one central and 48 ring emitters that are located circumferentially 13.5 m diameter. Antenna system's emitters are mounted on the reflector of 30 m diameter;
two controlling antennas of a near field;
and the remote controlling apparatus RCE-2000, which is situated on the command-dispatching position, being up to 10 km distant from the operating room.

RCE-2000 apparatus provides, if necessary, access to information about radio beacon parameters via local network Ethernet or via ATN network.

In radio beacon, all main devices are allocated. One package of all main radio beacon devices is located in the cabinet. An emergency package of devices with additional packages' switching devices is located in the second similar cabinet. In the process of work, the control of main and emergency packages' working capacity is provided. Switching onto an emergency package is performed automatically upon the control device's signal.

Table 3.1 Characteristics of azimuth radio beacons

Parameter	Parameter value	
	RMA-90	DVOR-2000
Operating zone: In horizontal plane In vertical plane (°)	From 0 to 360 3–40	From 0 to 360 0–40
By distance At 12000 m, km height At 6000 m, km height	Not less than 300 Not less than 100 (at 0.5 capacity)	Not less than 340 240
Contribution of a ground-based radio beacon into the error of azimuth determination, 2σ, not bigger (°)	±2	±1
Error of azimuth measurement in points at a distance from antenna's center (°)	Not bigger than 1 at a distance from antenna's center 28 m	Not bigger than 0.2 at a distance from antenna's center till 300 m
Carrier waves' frequency (MHz)	108.100–117.975	108,000–117.950
Deviation of carrier waves' frequency (%)	±0.002	±0.001
Reference phase's signal frequency (Hz)	30 ± 0.03	30 ± 0.02
Subcarrier's signal frequency (Hz)	9,960 ± 20	9,960 ± 10
Antenna system, diameter (m)	0.8	13.5
Antenna system reflector, diameter (m)	5	30

Electrical power of a radio beacon is provided by the main and emergency three-phase network 380/220 W, 50 Hz. During 30 min, radio beacon can work from batteries.

The beacon maintenance mode is uninterrupted, 24-hour, does not require constant staff assistance.

Functional diagram of radio beacon's signal forming apparatus is illustrated in Fig. 3.15.

The transmitter forms carrier frequency (CF) signal and four side frequency (SF) signals. Transmitter consists of:

exciter (transmitter controller; digital frequency synthesizer; SF signals former that consists of single-band modulators (4 items); CF amplitude modulator and SF amplitude modulators (4 items));
CF signal power amplifier;
SF signal power amplifier (4 items).

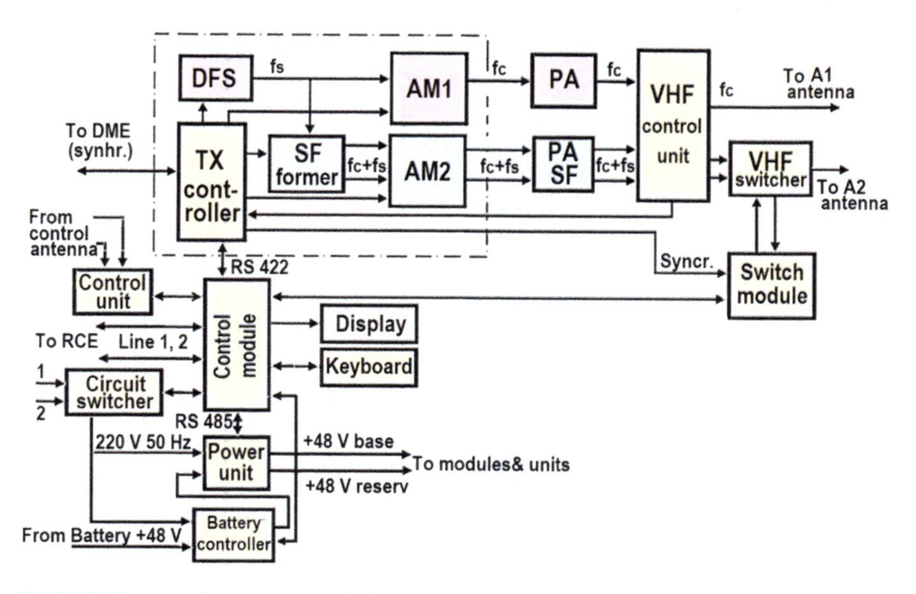

Fig. 3.15 Functional diagram of radio beacon's signal forming apparatus

Signal of carrier's frequency diapason 108.0–117.950 MHz, which is formed by digital frequency synthesizer (DFS), is transformed into AM signal with modulation frequency 30 Hz with the help of amplitude modulator (AM_1) and with the help of single-band modulators (SF former), it is transformed into the SF signals $f_0 \pm$ 9,960 Hz.

Transmitter forms four types of SF signals: frequency $f_0 + 9{,}960$ Hz with amplitude modulation by quadrature (sin and cos) frequency signals 30 Hz and frequency $f_0 - 9{,}960$ Hz also with amplitude modulation by quadrature (sin and cos) frequency signals 30 Hz.

Received CF signals of modulation frequency 30 Hz and SF that are modulated by quadrature (sin and cos) frequency signals 30 Hz amplify in the CF signals' power amplifier (PA and in four SF signals' power amplifiers).

CF signal from the output of power amplifier through the directional coupler enters the central emitter. After necessary transformations, one part of signal power from the second output of coupler enters the control module for signal's quality control.

SF signals coming from the power amplifiers through directional coupler enters the commutator (VHF switcher), which performs their pairwise connection to the corresponding opposite-side emitters, imitating their mechanical rotation.

In a commutator, continuous VHF signals $f_0 + 9{,}960$ Hz with sin and cos AM and $f_0 - 9{,}960$ Hz also with sin and cos AM are transformed into pulse ones and are then divided between 48 emitters of antenna array. Impulse VHF signals with AM are directed to even emitters according to sine law, while those modulated by the cosine law are directed to the uneven emitters. Switch module controls the switching process. Upon its signals, sequential connection of four transmitter outputs to the commuted emitters of antenna array is performed.

Emitters' switching in both—even and uneven group—is performed at the moment when the level of a signal's delivery power on the signal VHF inputs of a commutator is minimal. Thereat, even and uneven emitters are switched with a half-cycle offset (694 μs) by commutating pulses. Such type of commutation allows to create an effect of emitter's phase center smooth motion around the set circumference. VHF signal is "flowing" from one emitter to another. In the process of emission, two emitters are constantly used—even and uneven, if the power level is increasing in one of them, it leads to decrease in another one. Full circle of commutator's operation is 33.3 ms, and it is determined by reference and alternating phase's signals frequency (30 Hz).

From directed coupler's second outputs, SF signals enter the VHF signal control device. After necessary transformations, signal enters the control module from the control device's output. This signal is used for controlling and stabilizing the power output level of the corresponding transmitter's output.

In control device, with the help of frequency transformer, on the second input of which the carrying frequency enters, provides frequency of 9,960 Hz. This signal also enters the control module and is used to control SF signals' phase.

Transmitter control forms modulation signals considering parameters of transmitter's output signals and delivers the control signals on VHF modules.

CF and SF channels' signals must be phased; moreover, signals that are emitted by diametrically situated antenna array's radiators must as well be phased and equal in capacity. Otherwise, SF signal spectrum on board of aircraft will not be perceived as a whole. That is why the modulation control is performed by amplitude, as well as by VHF signal phase.

Transmitter's operation control and switching from one package to another and output signals' monitoring are performed by control module. Apart from that, control module serves the following functions:

forms supporting voltage-control circuits for transmitter;
performs commutation impulses' forming block, power module, and radio beacon's VHF passageway control;
determines radio beacon's parameters by signals from control antenna and transmits this information to RCE-2000 apparatus.

Control module consists of the following devices:

micro-PC controls all radio beacon's devices, using different types of interfaces. Micro-PC gets information about all devices' condition that is a part of radio beacon, analyses and forms substantive evaluation of its condition. Radio beacon's condition is displayed on indicators, "NORM," "DEGRADATION," "FAILURE" that are located on front panel of control module and is transmitted via communication line into the RCE-2000 apparatus. When detecting any changes in devices' condition ("DEGRADATION", "FAILURE"), sound alarm goes off.

router is meant to provide the information exchange between RCE-2000 and micro-PC apparatus;
demodulator is an analog–digital transformer which transforms signal from receiver's output into the digital form and passes it to the micro-PC;
indicator and keyboard provide ergonomic user interface for radio beacon controlling;
main controller—together with micro-PC, it serves as a generator of modulating signals and a device of CF channel operation monitor and control;
single-band controller is intended to form reference voltage for single-band modulators, mounted into the transmitter as well as to control the SF signals' phase using signals that come from VHF control devices.

Receiver P-872B performs acquisition, amplification, and detection of control antenna's signal, which then enters the control module for processing and main radio beacon parameters' control.

Control antenna (field inductor), receiver, and demodulator from the controlling module structure provide field monitor of radio beacon parameters control in near zone.

Amplitude and emitted signals' phase are settled so that a composite signal with required parameters is formed in space. This process is constantly traced using a monitor and transmitter controller.

Power supply module forms radio beacon devices' source voltage from the main or emergency network voltage ~220 V 50 Hz and performs control of input network's voltage and control of uninterruptive power supplies' operation.

Every transmitter has separate power supply. If one of the transmitters fails, the second one stays functional.

3.3 Distance-Measuring System

3.3.1 Function and General Characteristics of the System

Distance-measuring equipment (DME) is designed to determine a slant range of an aircraft relative to a ground-based radio beacon. The following tasks can be performed with the use of DME:

fixing the position of an aircraft using the intercept (LOP) method by measuring the distance to two ground DME beacons or, if collocated with VOR (DVOR), by measuring the distance and azimuth relative to the beacon. It should be noted that the accuracy of position fixing in these cases is higher than by measuring the azimuths to two VOR;
determining the ground speed of an aircraft based on the speed of slant range change and the time remaining before arrival to the destination if the DME information is used in the FMS computer;
flying precisely along the line of equal distances (arc of a circle);

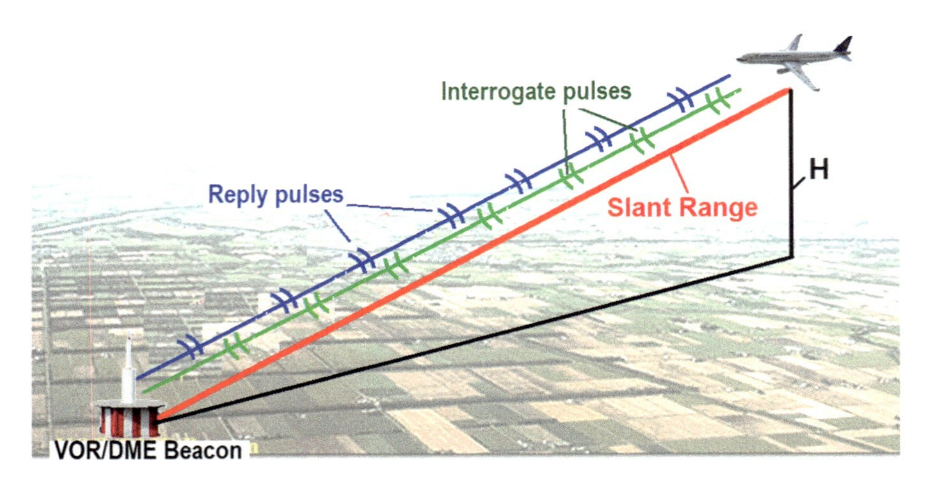

Fig. 3.16 Concerning the DME operation principle

determining and indicating accurate ranges to the runway threshold when approach an ILS/DME procedure.

In accordance with modern navigation technologies, DME is the basis for a simple area navigation (RNAV) system when the appropriate computerization is fitted. In doing so, DME provides accurate range inputs into the more complex and accurate RNAV systems; twin, self-selecting DME/DME are used.

On the principle of operation, DME is a pulsed distance-measuring system with a transponder. It includes a ground-based radio beacon (a transponder) and onboard equipment (an interrogator) (Fig. 3.16).

DME/N and DME/P systems are distinguished.

DME/N is a standard system used for navigating en route and in the terminal area. Ground-based DME/N beacons are usually installed as an augmentation to VOR in places with intense air traffic requiring higher navigation accuracy than VOR can ensure.

Due to this collocation, a VOR/DME short-range navigation system is formed. The VOR and DME antennas can either be located on the same vertical axis (coaxial collocation) or be spaced (at most 30 m—for flights in the terminal area, at most 600 m—for flights along the airways).

The DME beacon can also be used jointly with the beacons of landing systems (ILS or MLS). In this case, it is installed at the airdrome, so that the distance to it, measured by the airborne equipment, is equal to zero in the touchdown zone.

In the joint use of DME beacons, their frequencies are paired to VOR or ILS frequencies, and a DME interrogator is designed to tune automatically to the corresponding DME frequency when the associated VOR or ILS frequency is selected.

DME/P ("P"—precision) system differs from DME/N in increased accuracy of measuring a distance. Originally, DME/P was developed for MLS landing system. Nowadays, DME/P radio beacons are used in ILS system as well.

DME/P has two operational modes—initial approach (IA) and final approach (FA) with different parameters of signals of the interrogator and transponder.

Through an omnidirectional antenna, the DME beacon forms and radiates vertically polarized signals, P0N emission class. DME interrogation and reply signals are pulse pairs radiated at random intervals, i.e., the transmission sequence of pulses is irregular or jittered.

The onboard equipment includes an interrogator and an antenna providing omnidirectional reception of signals. Two sets of equipment can be installed on an aircraft with the purpose of improving the reliability and making it possible to tune to two different DME beacons, thus providing aircraft position fixing with the use of intercept (LOP) method. However, there already exist interrogators which can measure ranges to several beacons sequentially (up to five, for example, SD-75M, Russia).

To indicate the measured range to the beacon, different indicators on different types of aircraft are used: multifunction PFD indicators (Fig. 3.17a), ND (Fig. 3.17b), or separate indicators (Fig. 3.17c).

DME works in the frequency range of 960–1,215 MHz at 1 MHz spacing; this provides 252 spot frequencies (frequency and code channels). The channels are numbered 1–126X and 1–126Y.

The channels of interrogation (air-ground, the band of 1,041–1,150 MHz) and of transponding (ground-air, the band of 962–1,024 MHz and 1,151–1,213 MHz) are frequency-spaced. There is always a difference of ±63 MHz between the interrogation and transponding frequencies.

Frequency spacing of interrogation and transponding channels is necessary to eliminate false activation of transmitters by reply echo signals reflected from different objects in the DME installation area. For the same purpose, coded signals consisting of two pulses are used for interrogation and transponding. They are coded by setting different time intervals between the pulses of interrogation and beacon reply signals. The use of several interrogations and reply codes at each carrier frequency also provides an opportunity to increase the number of operating frequency and code channels.

The time interval (code) between the pulses of interrogation signals is 12 μs or 36 μs. The time interval (code) between the pulses of reply signals is 12 μs for the band of 962–1,024 MHz and 30 μs for the frequency band of 1,151–1,213 MHz.

A quite small frequency spacing between adjacent channels of interrogation and transponding (1 MHz) can lead to an influence of side lobes of the pulse signal spectrum on adjacent frequency channels. To eliminate this, DME pulse signals have a special bell-like shape (Fig. 3.18) as well as a relatively large duration (3.5 μs at level 0.5 of the pulse amplitude) and a large duration of leading and trailing edges (3–3.5 μs). These parameters of the pulse envelope provide maximal radiated power for the minimal width of the signal spectrum and the off-frequency emissions level.

Fig. 3.17 Kinds of indicators for presenting DME information

The area of DME coverage is determined by the line-of-sight range and depends on the transponder power. The typical range is 360 km en route and 100 km in the airport area.

The mean pulse power of the beacon transmitter depends on the frequency of generating reply signals which, in its turn, is determined by the frequency of coming interrogations. Such operation of the transmitter causes changes in its output power which has an adverse impact on the operation of its output cascades.

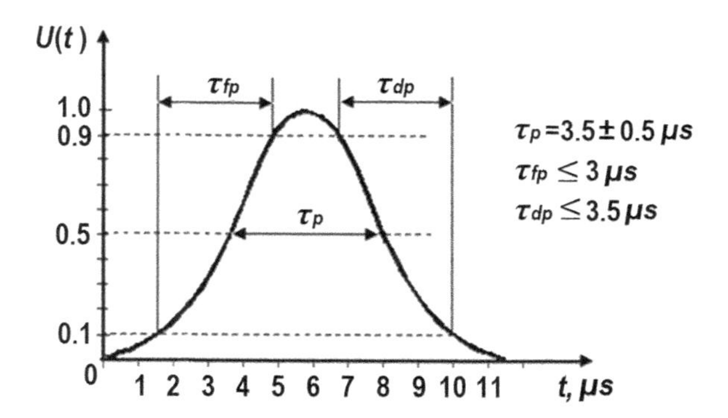

Fig. 3.18 Envelope of a DME pulse

To improve the transmitter operation, the beacon continuously emits coded pulse pairs with random timing even if there are no interrogation signals. These pulse pairs are squitter pulses (SP) with a repetition frequency of at least 700 pp/s (pp/s = pulse pairs per second). These signals are structurally similar to reply signals, but the moments of their generation (and the periods) are random, since they are determined by intersecting the set threshold level by receiver noises.

The shape and parameters of the envelope of radio pulses are identical for SP, interrogation, and reply signals.

For identifying, a DME beacon transmits an identification signal. The identification signal is a Morse code signal which corresponds to the three-letter code assigned to the beacon. The identification signal is radiated by the beacon every 40 s and has duration of 10 s and speed at least six words per minute. In doing so, the radiation of the range reply signal and SP is interrupted.

For tonal filling of dashes and dots corresponding to Morse letters, coded pairs of range reply are used and they are generated with the frequency of 1,350 Hz. The DME audio signal is processed by the interrogator and sent to the AMU and can be heard by the crew on headphones or cockpit loudspeaker in parallel. As tonal filling, other frequencies can be generated as well.

If collocated with a VOR or ILS, it will have the same identity code as the parent facility. Additionally, the DME will identify itself between those of the parent facility. The DME identity is 1,350 Hz to differentiate itself from the 1,020 Hz tone of the VOR or the ILS localizer.

Map symbols of a single DME as well as DME collocated with VOR used in aeronautical charts are shown in Fig. 3.19. You can also see here how DME beacons are displayed on a multifunction ND of a modern aircraft.

There is letter “D” preceding the frequency value 113.4 in the box on the chart. It indicates DME/VOR collocation. The DME frequency is not stated because when collocating with a VOR or ILS localizer, and it is determined by the frequencies of these beacons.

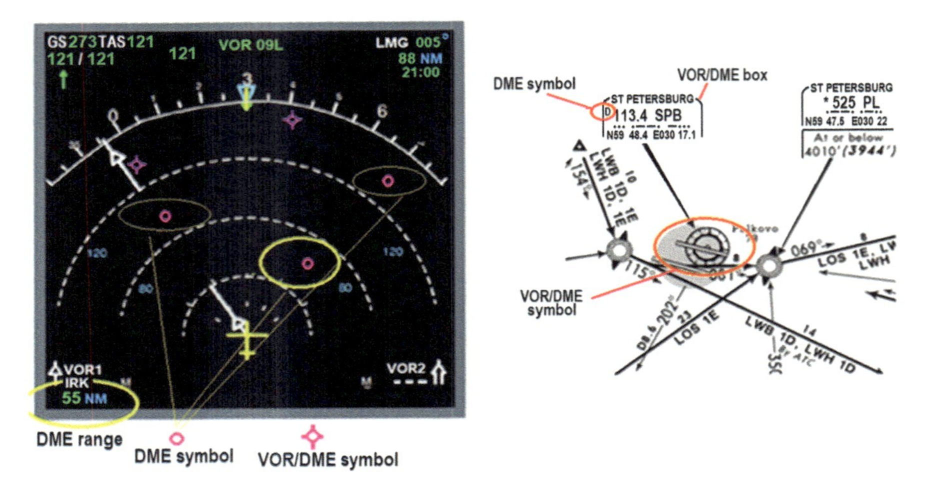

Fig. 3.19 Chart and ND symbols of the beacon

3.3.2 DME Generic Block Diagram and Operation Principle

In DME, the range is determined by a delay between the time of radiating an interrogation signal by the onboard equipment and the time of receiving a reply signal from the ground-based beacon.

A total delay between the time of radiating an interrogation signal and that of receiving a reply signal is determined by equation:

$$\tau = 2R/c + t_{\text{del}},$$

where R is a distance between the interrogator and transponder, t_{del} is a signal delay in the transponder equipment. Then, the measured range is determined as:

$$R = \frac{c}{2}(\tau - t_{\text{del}}).$$

The signal delay t_{del} in the transponder formed from the time needed for processing the interrogation signal received. But DME contains an intentionally introduced time delay t_{del}, so that the minimal value of a measured range is equal to zero. Indeed, if $t_{\text{del}} = 0$, at small ranges, the reply signal enters the receiver of the interrogator still during radiating the interrogation signal when the receiver is blocked. This leads to the occurrence of silent zone of the pulsed DME with dimensions determined by minimal measurable range:

$$R_{\min} = \frac{c}{2}\left(\tau_p - t_{\text{del}}\right),$$

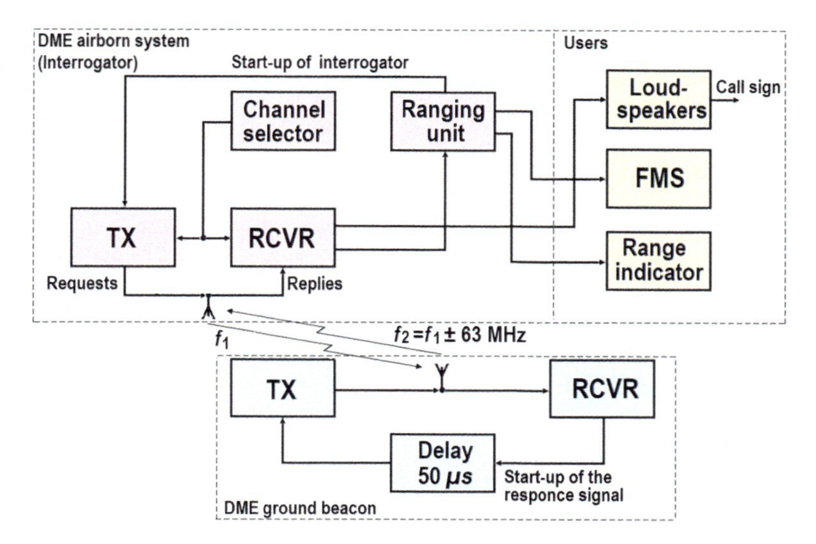

Fig. 3.20 Simplified block diagram of the DME system

where τ_p is duration of the interrogation signal.

To ensure a minimal measurable range equal to zero, the time of radiating a reply signal by DME should be delayed in respect to the time of receiving the interrogation signal for the interval not less than the duration of the interrogation signal.

Periodicity of radiating the interrogation pulses can lead to unambiguity of measuring the distance when the subsequent interrogation signal is radiated before entering the reply signal to the previous interrogation.

To eliminate the unambiguity, the period of repetition of interrogation pulses should exceed the maximal total delay when propagating a signal from the interrogator to the transponder and back for the maximal coverage range $R_{\max}$ of the system, i.e., the condition:

$$T_{\text{rep}} \geq \frac{2R_{\max}}{c} + t_{\text{del}},$$

must be met.

If we accept that $R_{\max} = 450$ km, $t_{\text{del}} = 50$ μs, then $T_{\text{rep}} \approx 3 \times 10^{-3}$ s. Hence, the frequency of interrogation pulse repetition must be at most 330 Hz.

A simplified block diagram of the DME system is shown in Fig. 3.20.

The scheme of range computation is a digital meter of the interval between the time of radiating an interrogation signal and that of receiving a reply signal. The interrogator coder is designed to form a two-pulse interrogation signal with a code chosen by the channel selector. The interrogator decoder is designed to form a pulse from a two-pulse reply signal. The time of receiving the reply signal is fixed on the leading edge of the pulse.

The calculated range value is indicated and enters the FMS computer. The aural signal of the beacon identification extracted in the DME receiver is sent to the pilots' headphones to be heard.

The peculiarity of DME operation is that the beacon generates reply signals (of SP-type) regardless of the presence or absence of interrogation signals from onboard interrogators (squitter pulses and replies to requests of aircrafts). According to ICAO requirements, the average frequency of SP signal generation is at least 700 pp/s.

When receiving interrogation signals, the beacon generates squitter pulses and replies to requests of aircrafts. The maximal frequency of generating such summarized reply signals must be at least 2,700 pp/s. The higher the frequency of entering the interrogation signals is, the lower the frequency of generating the SP replies is.

The value of 2,700 pp/s as the maximal frequency of replies is based on the following reasons. A peculiarity and yet a main disadvantage of the DME system as a system with a transponder is its limited capacity defined by the number of aircraft served by the system per unit of time or simultaneously. For DME, the capacity is 100 aircraft per second with reply efficiency $K_{rep} = 0.9$ defined. The reply efficiency is the ratio of replies to coming interrogations.

With an average frequency of interrogations from the onboard interrogator equal to 30 Hz, the number of coming interrogations from 100 aircraft is 3,000 per second. Then, with the reply efficiency $K_{rep} = 0.9$, the beacon must generate 2,700 pp/s.

There are beacons/transponders with a frequency of generating the reply pulse pairs different from the frequency mentioned above. So, for example, DME FSD-45 beacon has a minimal frequency of transmitting pulses of 800 pp/s and a maximal one of 3,800 pp/s with the reply efficiency at least 0.7 (minimum allowed value).

When defining a maximal frequency of generating the replies, the key criterion is providing the capacity of at least 100 aircrafts with the given reply efficiency.

It should be also noted that the transponder receiver gets blocked for a so-called dead time (usually not less than 60 μs) after receiving interrogation pulses. This time is a measure of time required for processing (decoding) of interrogation signals in the beacon equipment. The blocking of the receiver eliminates the influence of signals re-reflected from ground features (echo signals) which are perceived as interrogation ones.

If there are no interrogation signals at the receiver input, the sensitivity of the receiving device of the beacon is maximal.

When the ground-based beacon receives range interrogation signals from an aircraft, it generates one group of replies for the interrogation signals, the other—for dealing with noise. If the number of interrogations from the aircraft within the beacon coverage range exceeds 2,700 pp/s, the sensitivity of its receiver decreases and the aircraft flying at large distances from it are not served (their interrogation signals get lower than the detection threshold). Thus, the total frequency of generated reply pulses and SP does not exceed 2,700 pp/s (or the maximal number of replies per second specified for the given beacon type).

After tuning to a frequency and code channel of the beacon, the onboard DME equipment operates in the mode of searching any random signals of the beacon.

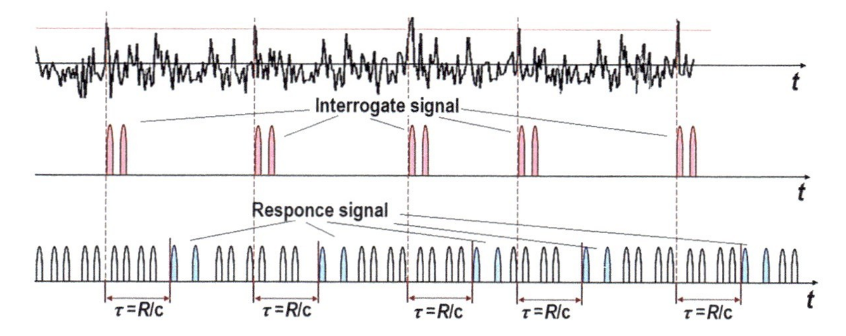

Fig. 3.21 Concerning the principle of picking up intended reply pulses

After receiving an SP of the beacon, the airborne DME automatically starts to generate interrogation signals in the form of coded pulse pairs for the beacon, within the coverage range of which it is situated. The period of pulse pair repetition is random, thus increasing the system immunity resistance to synchronous pulse interference.

After receiving and decoding an interrogation signal from the onboard DME, generation of a coded pulse pair starts. After a delay for a fixed time interval of 50 μs (for *X*-code channels) or 56 μs (for *Y*-code channels) in the beacon equipment and giving the bell shape to the envelope, these pulses modulate a signal of the carrier generated by the UHF generator of the transponder transmitter and are radiated as reply signals.

At the input of the interrogator receiver, there exist both SP reply signals and intended replies to its own interrogations with the time of their coming being unknown and requiring determination. When searching reply pulses for own interrogations, the frequency of radiating the interrogation signals is maximal and equal to 150 Hz (150 pp/s). After 100 s, it reduces to 60 Hz and remains at this level till intercepting intended reply signals or till tuning the onboard equipment to a new frequency.

To pick up intended reply signals, the onboard equipment uses time gating (Fig. 3.21). It is based on the fact that the delay of a reply signal for an own interrogation does not depend on the time of radiating the interrogation signal and is determined only by range to the beacon which either is constant or changes slowly in time between adjacent interrogations. So, if the tracking pulse gate is shifted relative to the time of generating an interrogation signal for an interval determined by the range to the beacon plus a fixed delay of 50 (56) μs, the intended reply signals will fall within its duration. The decision about receiving intended replies is made when a definite number of reply pulse pairs fall within the pulse gate for a given time interval.

In the modern equipment, the time of searching and intercepting intended reply pulses does not exceed 1 s. For this time, it is required to receive in average 27 reply pulse pairs falling within the pulse gate.

When receiving the beacon reply pulses synchronous with the interrogations, the onboard equipment goes into a tracking mode—a mode of measuring the distance.

In the tracking mode, the frequency of radiating the interrogation pulses reduces to 30 Hz.

Possessing an increased accuracy of measuring a distance, DME-P has some differences from DME-N both in signal structure and in radiation mode.

The accuracy of measuring a distance with the use of pulse method depends on steepness of the leading edge of pulses. So, in DME-P, steepness of the leading edge of pulses increases at small distances to the beacon (at final approach). At the same time, however, the width of the signal spectrum increases, which requires an appropriate widening of the pass band of the receiver paths. Without adopting additional measures, this would require decrease in number of frequency and code channels of the system.

In order that the system can retain 252 frequency and code channels, DME-P has some additional measures introduced, and namely two more code distances between pulse pairs (Z and W codes) and initial approach (IA) and final approach (FA) operation modes. The FA mode is characterized by a higher accuracy of measuring a distance than the IA mode.

Transition from IA to FA is accomplished automatically at the distance of 13 km to the beacon. The IA mode uses one code distance between interrogation pulses, the FA mode uses another one.

When operating with DME/P beacons, the interrogation signals have the following codes:

IA mode $X = 12\ \mu\text{s}$, $Y = 36\ \mu\text{s}$, $Z = 21\ \mu\text{s}$, $W = 24\ \mu\text{s}$;
FA mode $X = 18\ \mu\text{s}$, $Y = 42\ \mu\text{s}$, $Z = 27\ \mu\text{s}$, $W = 24\ \mu\text{s}$.

As for reply signals:

$X = 12\ \mu\text{s}$, $Y = 30\ \mu\text{s}$, $Z = 15\ \mu\text{s}$, $W = 24\ \mu\text{s}$.

In DME/P, the frequency of interrogation pulse repetition is also different from DME/N and equals to 40 pp/s in the mode of searching reply signals, 16 pp/s—at IA stage, 40 pp/s—at FA stage and 5 pp/s when the aircraft is on the ground (fulfilling land operations).

For DME/P, the maximal average frequency of generating the reply signals is about 1,200 pp/s.

Application of DME/P in the MLS or ILS landing system provides for the following interaction technology of the interrogator and the beacon. At a distance of about 40 km from the beacon, the interrogator radiates signals with codes and pulses of IA mode. At a distance of 14–15 km, the interrogator starts radiating signals with codes and pulses of FA mode. If there is a beacon of the same standard at the airdrome, the two-way FA mode is activated from a distance of 13 km.

Fig. 3.22 Appearance of DME beacons and antennas

3.3.3 Design and Main Specifications of DME Beacons

As a rule, DME beacons are collocated with VOR (DVOR) (Fig. 3.14) but a separate installation either in the operating room (a container) or in a building is also possible (Fig. 3.22).

The beacon antenna system consists of a transceiving and monitoring antennas combined structurally. The transceiving antenna can contain vertical rows of half-wave dipoles located at generatrices of a cylinder with a diameter of about 15 cm. The dipoles are fastened on a metal structure serving as a reflector and are closed with a shared fairing with a diameter of 20 cm. If DME is collocated with VOR, the DME antenna is mounted over the VOR antenna.

The beacon antenna provides omnidirectional radiation and reception in horizontal plane. The antenna pattern in vertical plane is described by cosecant function and provides, for the sector of angles between 0 and 40°, an excess of a primary signal over a reflected one with any underlying surface: from water to tight soil.

The main specifications of DME beacon DME-2000 produced in Russia are shown in Table 3.2.

Consider peculiarities of the DME design and operation using the example of DME-2000 (Russia).

DME-2000 is designed for using en route and at the airports and operates in DME/N (DME/P) signal formats. The beacon can be used jointly with VOR (DVOR) beacons, ILS independently.

The beacon includes a transceiver stand (box), a transceiving antenna, and remote control equipment RCE-2000.

The beacon is controlled via the panel of local control or RCE-2000 which can be located at a distance of 10 km. Changes in the equipment state or beacon parameters are accompanied by light and audible indication. The RCE-2000 equipment provides,

Table 3.2 Specifications of DME beacon

Parameter	DME 2000
Coverage area in horizontal plane (°)	0–360
Coverage area in vertical plane (°)	0–40
Coverage area by range (under direct visibility):	
At the flight altitude of 12000 m, at least (km)	≥340
At the flight altitude of 6000 m, at least (km)	≥240
Error introduced by the beacon in measuring a distance, 2σ, m, at most:	
When interacting with VOR, DVOR (m)	±150
When interacting with ILS	±75
Number of aircraft served simultaneously	200
Frequency band (MHz)	962–1,213
Pulse shape and other parameters	According to ICAO requirements
SP frequency (pp/s)	700
Frequency of generating a range reply, pp/s, at least	2,700
Automatic change-over from a failed set to a standby one for at most 3 s when:	
Changing the delay time (μs)	±0,5
Changing the code interval (μs)	±1,0
Failing the control device	+
Time of operation with an emergency source (min)	≥30

if necessary, Ethernet- or ATN-access to the information about the beacon parameters condition.

The built-in test system provides automatic control of all main parameters of the beacon, search and isolation of faults within the accuracy of a replaceable module (board). The beacon parameters and equipment state are indicated on a color display in graphics mode. All changes in the equipment state as well as maintenance personnel's actions are documented and saved for 30 days in the RCE-2000 equipment.

To increase the reliability, all main devices are backed up in the beacon. The main and standby sets of equipment are located at the same stand of the transceiver. When operating, the main and standby sets are monitored for performance capabilities. Change-over to the standby set is performed automatically on signal from the controlling device.

The electrical power for the beacon is supplied from the main and emergency network of 220 V, 50 Hz. The beacon can operate on batteries for 30 min. The beacon is meant for continuous round-the-clock operation and does not require the continuous presence of maintenance modules, boards, and devices as well as modern element base and surface-mount technology are applied in the beacon equipment. There are two digital signal processors with a timing frequency of 300 MHz in the

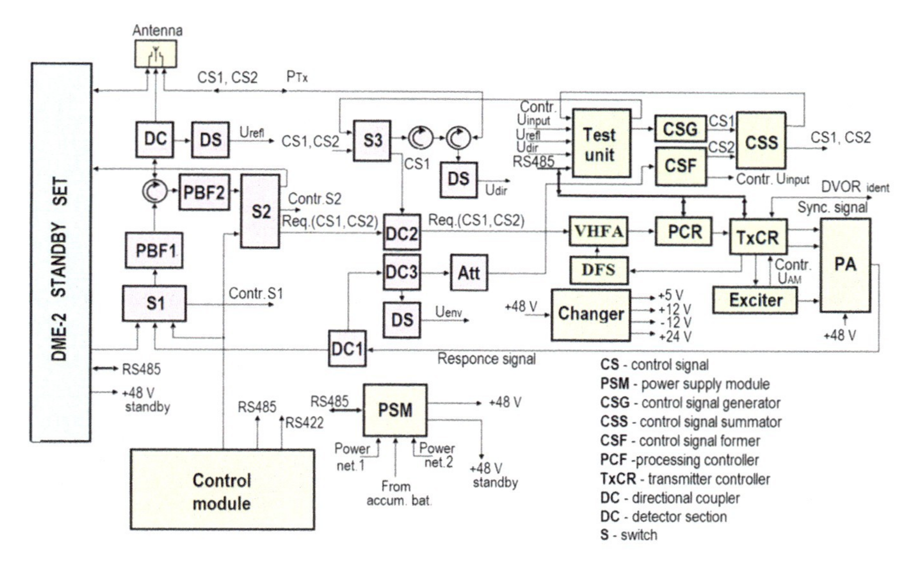

Fig. 3.23 Functional diagram of DME-2000

receiving and processing channel. The digital processing of input signals is performed directly on the intermediate frequency.

The functional diagram of DME-2000 is shown in Fig. 3.23.

The HF switch connects reply pulses from the main or standby set of the transponder to the antenna (through switch S1). Switch S1 is controlled by the signals from the control module.

The control module also provides connection of the antenna output—interrogation signals (through switch S2) to the HF module of the operating set.

Cross-coupling of the reception and transmission channels in the HF switch is provided by the circulator. The directional coupler (DC) and detector section (DS) ensure control of coordination of antenna impedance with the transmitter output by level of the reflected signal.

The interrogation signal enters the processing module through the directional coupler (DC2) of the HF module. Here, the signal is amplified and converted to the intermediate frequency (a heterodyne signal is generated by the digital synthesizer of the receiver frequency).

The function of the processing controller is detection, delay of generating a reply signal for a specified time, generation of a coded reply signal. The transmitter controller controls the choice of code parameters and carrier frequency formed by the exciter. The power amplifier ensures carrier modulation by a pair of coded pulses.

The processing module controls parameters of the reply signals generated by the beacon (with the aid of the controlling device) and generates control signals (with the aid of the generator and former of control signals) designed to test the functional efficiency of the beacon modules.

3.4 Summary

Short-range navigation systems, which use VOR and DME radio beacons as a basis, showed its high efficiency at the flight navigation's tasks solving. They have quite high accuracy, functionality, and good operational characteristics. Thus, it is possible to change the short-range navigation system's configuration by using different combinations of radio beacons: VOR/DME, DME/DME, or VOR/VOR. Specified advantages as well as the practice of usage, allow VOR/DME system to be considered the main radio navigation system of civil aviation at this time. One can add that in the military aviation TACAN (USA) and RSBN (Russia) systems are also considered the main aid of aircrafts' radio navigation of tactical radius of action.

One of the main disadvantages of such systems can be considered the operating range which is limited by straight visibility. This leads to the necessity of using the network consisted of many radio beacons in order to create the continuous navigation field that requires significant financial expenses. In some parts of the world and states which are characterized by the advanced ground infrastructure of navigation aids, this task is solved. It allowed to increase the efficiency of airspace use in such regions by means of implementing modern navigation technologies, which are based on the PBN concept. Thus, lots of navigation technologies and applied navigation processes (according to the RNAV and RNP navigation specifications) are based on the use of VOR/DME and DME/DME systems.

That is why we can suppose that VOR/DME system will continue to develop and be used in aviation in the mid-term perspective despite of GNSS system development.

3.5 Further Reading

Additional information on VOR and DME system, its main characteristics, and operating principles can be found in [1].

Information on construction, operating principles as well as on the main technical characteristics of Russian-made onboard and ground aids VOR/DME can be found in [2, 3].

Information on ICAO requirements to the parameters and characteristics of VOR/DME radio beacons can be found in [4].

There is a wealth of information on VOR at https://en.wikipedia.org/wiki/VHF_omnidirectional_range.

There is a wealth of information on DME at https://en.wikipedia.org/wiki/Distance_measuring_equipment.

References

1. Kayton M, Fried WR (1997) Avionics navigation systems, 2nd edn. A Wiley-Interscience Publication (USA), Wiley
2. Skrypnik ON (2014) Radionavigacionnye sistemy vozdushnyh sudov [Radionavigation systems of aircrafts]. Moscow, INFRA-M (in Russian)
3. Yu M, Tobolov, K (2013) Azimutal'no-dal'nomernaya sistema VOR/DME. Postroenie i ehkspluataciya. [Azimuth and distance system VOR/DME. The construction and operation]. Siberian Branch of the Institute of Air Navigation. E-learning course (in Russian)
4. Annex 10 to the convention on international civil aviation. In: Aeronautical telecommunications, vol 1, 6th edn. Radio Navigation Aids (2006)

Chapter 4
Radio-technical Landing Systems

Landing is the most complicated stage of flight at the end of which an aircraft should be taken to the set point on the RW surface. Different technical aids which form the landing system are used in order to solve this task. Landing systems have strict requirements which are connected with the provisioning of predefined security level of this stage of flight. Radio-technical landing systems satisfy these requirements the most.

First landing systems which used radio signals were created in the beginning of 1930s, and they showed its efficiency. During the following years, scientific researches and events were conducted which were focused on the enhancement of performance and technical characteristics of radio-technical landing systems.

Several types of radio-technical landing systems have found its use in practice: those based on radio beacons NDB, landing radio locators, special landing (course and glide slope) radio beacons. The best characteristics of accuracy and reliability of putting the aircraft into the tangency point RW have instrumental landing system, which backbones are localizer and glide slope radio beacons.

Currently, airfields that are characterized by mid- and high intensity of air traffic are equipped with these landing systems. The use of the most advanced systems of this type provides the required flight safety and high capacity of airfields in simple and complicated meteorological conditions, during the day and night. That is why this chapter gives special attention to ILS.

Section 4.1 shows the classification of landing systems, categories of landing minimum are described, and special terminology is explained which is used for landing system characterization.

Section 4.2 provides characteristics of simplified, radio-locating, radio-beaconing landing systems as well as their content, positioning on the airfield and operating principle. Their advantages and disadvantages are analyzed.

Section 4.3 describes instrumental landing systems: principles of construction and functioning, methods (equisignal and CSB/SBO) of setting the directions in space, and block diagrams of radio beacons which implement these methods. Mathematical equations are shown which describe the processes of forming the fields of emission of localizer's and glide slope's equisignal and CSB/SBO types. Influence of the

O. N. Skrypnik, *Radio Navigation Systems for Airports and Airways*,
Springer Aerospace Technology, https://doi.org/10.1007/978-981-13-7201-8_4

earth surface on the direction diagrams of radio beacon's antenna systems is shown. Constructional and operating principles of two-channel radio beacons are described.

In the last paragraph, processing of radio beacon signals in the onboard equipment of landing systems of analog and digital types is described.

Section 4.4 describes index radio beacons and radio receivers that are included in the ILS-type landing systems: main characteristics, simplified block diagrams of radio beacon and receiver, and signals' diagrams.

Section 4.5 describes the special aspects of construction and main characteristics of ILS radio beacons. As an example of realization, Russian-made radio beacons of landing system ILS 2700 are overviewed: construction and main characteristics Loc2700 and G/P 2700, their operating principles, and simplified functional diagrams.

4.1 Classification and Categories of Landing Systems

Landing is the most challenging flight stage after which the aircraft must be precisely targeting to runway—a rather limited land area. Landing is greatly influenced by meteorological factors, density, and intensity of air traffic in the vicinity of airdrome.

While landing, the aircraft height and speed considerably change, and maneuvers are executed in close proximity to the ground. This leads to the increase in the psychological load on the crew and to the rise of probability of action errors. At the same time, the opportunity to correct the errors of the crew or automatic systems decreases with the reduction of height. According to the ICAO's statistics, it is the landing stage which involves the highest percentage of aircraft accidents (disasters).

Approach and landing can be generally considered as a dynamic guidance of an aircraft into a region of admissible lateral and vertical deviations from the desired flight path at the decision height with the required probability. Getting to the region ensures (provided the speed of the aircraft is within the limits) execution of the necessary correcting maneuver and landing in the desired zone of the runway.

***Decision Height* (*DH*)** is the height of a flying aircraft above the ground at which the crew makes a decision about landing or going around (or diverting to another airdrome). Actually, this is a height of transition from instrument to visual flight.

To provide high efficiency of aviation, landing must be accomplished on the first try, in any meteorological conditions, day and night. For this purpose, different airborne and ground-based technical navigational facilities constituting a landing system are used.

The basis of a landing system is its ground-based part which may contain different technical facilities, first of all, radio-electronic and lighting equipment. It is designed to guide an aircraft to the terminal area, manage the air traffic in the area, provide landing, and manage the aircraft movements (taxing) on the airfield.

The landing system must provide the aircraft flight to the place of landing along a specified trajectory (a glide path) and its landing. The glide path is made up by intersecting the course and descending (gliding) surfaces (Fig. 4.1). The course surface

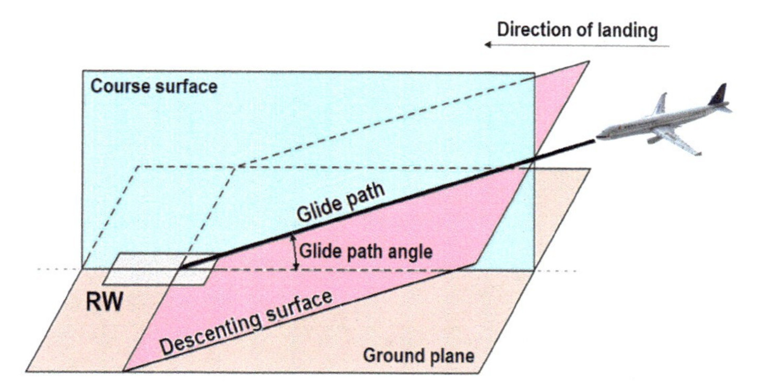

Fig. 4.1 Principle of glide path formation

is a vertical plane passing through the runway centerline. The descending (gliding) surface is a plane which is sloped at a specified angle to the runway surface and touches it at a certain distance from its threshold.

The landing system must define deviation from the glide path and the distance to the runway threshold continuously or at fixed points. The glide path can be set either really (e.g. using radio signals) or virtually (e.g. using FMS computations for the airdrome).

The landing conditions are characterized by values of the parameters constituting the so-called ***landing minimum***. The parameters include the minimal ***Runway Visual Range*** (***RVR***) (or horizontal visibility) and ***DH*** (cloud base height) which are directly related to the weather conditions in the vicinity of the runway. They are such distance to the runway and such flying height which provide reliable visibility of the ground and necessary runway orienting points for the crew.

According to the ICAO standards, Categories I, II, and III (A, B, and C) of landing minima are distinguished (Fig. 4.2):

Category I (***CAT I***) operation. A precision instrument approach and landing with a DH not lower than 60 m (200 ft) and with either a visibility not less than 800 m or a RVR not less than 550 m.

Category II (***CAT II***) operation. A precision instrument approach and landing with a DH lower than 60 m (200 ft), but not lower than 30 m (100 ft) and a RVR not less than 300 m.

Category IIIA (***CAT IIIA***) operation. A precision instrument approach and landing with a DH lower than 30 m (100 ft) or no DH and a RVR not less than 175 m.

Category IIIB (***CAT IIIB***) operation. A precision instrument approach and landing with a DH lower than 15 m (50 ft), or no DH and a RVR less than 175 m but not less than 50 m.

Category IIIC (***CAT IIIC***) operation. A precision instrument approach and landing with no DH and no RVR limitations.

The following types of landing minima are distinguished:

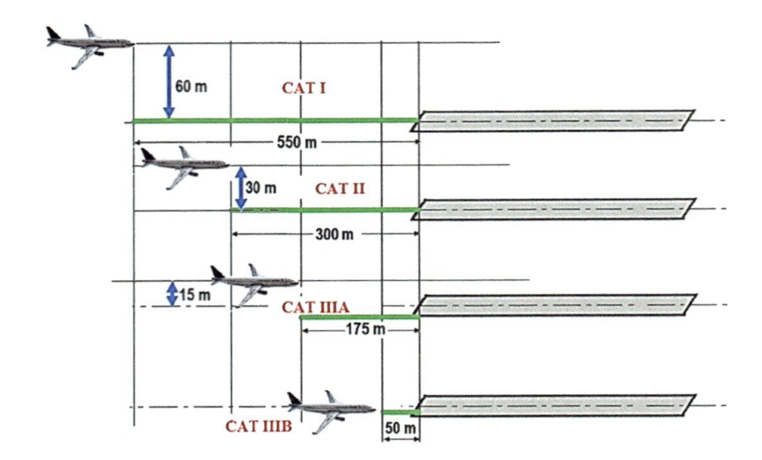

Fig. 4.2 Landing minima

that of an airdrome (is defined by characteristics of ground equipment of the landing system);
that of an aircraft (is defined by characteristics of airborne equipment of the landing system);
that of a crew (is defined by skills of the crew, first of all, of the captain).

In every particular case, the landing minimum is defined on the basis of minima of the airdrome, aircraft, and crew and depends on the highest of them.

The ***visual (non-precision) approach*** places higher requirements to the minima of the airdrome, aircraft, and captain. They are calculated with the aid of special techniques taking into consideration the airborne and airdrome radio navigation facilities, existence of natural and artificial obstacles in the vicinity of the airdrome, landing speed, and aerodynamic performances of the aircraft.

The values of the weather minima are contained in the aeronautical information publications (AIP) for each specific airdrome and specific aircraft class.

The typical values of the weather minima are:

NDB approach: RVR = 1,500–2,000 m, DH = 110–130 m;
visual approach for class IV airplanes (single-engined airplanes like Cessna and other similar planes) and all helicopters: RVR = 2,000–3,000 m, DH = 150–300 m;
visual approach for class I, II, III airplanes (twin-engined airplanes like ATR-72, SSJ-100 and all heavier ones): RVR = 5,000 m, DH = 600 m.

One of the major changes forecast for approach-to-landing phases is the increased use of RNAV and RNP to achieve optimum airspace utilization and noise abatement. The use of RNAV and RNP for departure procedures will allow increased flexibility in departure procedure design and will increase the ability of procedures to avoid noise-sensitive areas. GNSS use is fundamental for approaches according to the

RNP specifications (RNP 1, A-RNP, RNP APCH, RNP AR APCH, RNP 0.3) in correspondence with the PBN concept.

The landing systems are classified depending on the equipment and the way of determining the aircraft position relative to the glide path. The following types of radio-technical landing systems are distinguished:

Radar landing system (RLS)—the aircraft position relative to the glide path and to the touchdown zone is determined on board the aircraft using a controller's instructions on the basis of information from ground radars.
NDB-based landing systems—the aircraft position relative to the glide path and to the touchdown zone is determined using the signals received on board the aircraft from a ground-based NDB (locators) and a marker as well as an airborne low-range altimeter (LRA);
Radio beacon landing system (of VHF, UHF and SHF bands)—the aircraft position relative to the glide path is determined using the signals received on board the aircraft from ground-based radio beacons—localizer (LOC) and glide slope (G/S), and relative to the touchdown zone using the signals of a DME transponder and a marker. The information about the aircraft position relative to the glide path is displayed on special indicators in the cockpit.

According to the type of the used landing system, instrument and non-instrument approaches are distinguished.

Non-instrument approaches are visual approaches and visual flight rules (VFR) approaches. The crew determines aircraft attitude and position visually using the natural horizon, landmarks as well as other objects and buildings.
Instrument approaches are divided into precision approaches (suit ICAO categories) and non-precision approaches and are performed under instrument flight rules (IFR).
Precision approaches provide the aircraft with lateral and vertical guidance at the final stage (on the glide path). The radio beacon landing system, RLS, and RLS in conjunction with NDB-based landing systems are used during precision approaches.
Non-precision approaches provide the aircraft only with lateral guidance at the final stage (on the glide path). NDB-based landing systems, single NDBs, combination of single NDBs and the airport surveillance radar (ASR), VOR/DME system, and GNSS satellite navigation system are used during non-precision approaches.

4.2 General Description of Radio-technical Landing Systems

4.2.1 NDB-Based Landing Systems

NDB-based or simplified landing systems (*Oborudovanie Sistemy Posadki (OSP)—in Russian*) provide guidance of an aircraft to the terminal area, fulfillment of the final (lineup) maneuver, putting the aircraft on the final course as well

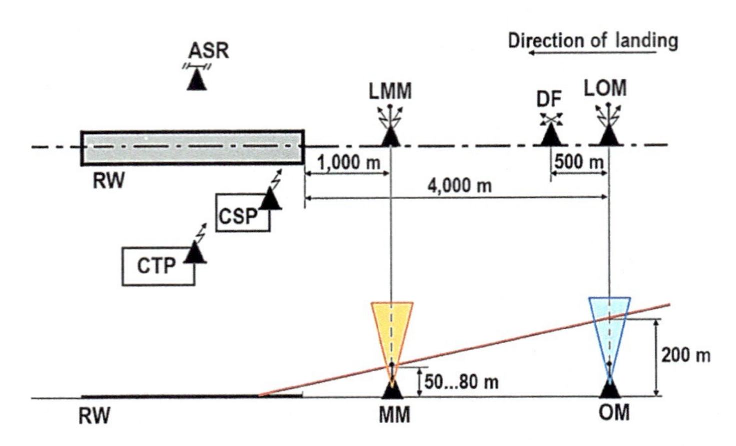

Fig. 4.3 Placement of the ground equipment of a simplified landing system (OSP)

as a descent check in two fixed points of the landing trajectory. At the final stage, the landing (final approach) is performed visually. OSP approach is called ***2NDB approach***.

The OSP include:

ground equipment: 2 NDBs (far and near beacons) with marker beacons (compass locator and outer marker (LOM), compass locator and middle marker (LMM)), a DF, an ASR, airdrome lighting equipment, and a VHF radio station (Fig. 4.3);
airborne equipment: a marker receiver, an ADF, a LRA, and a VHF radio station.

The LOM serves to guide an aircraft equipped with the ADF to the terminal area and to construct the final maneuver in the terminal area.

The LMM is designed to maintain a given flight direction determined by the final course after the aircraft passes over the LOM. On passing the LOM, the ADF automatically switches over to the LMM operating frequency.

The outer marker (OM) allows the crew to check the height of flight, distance to the touchdown zone, and readiness of onboard systems to provide flight at the final stage of approach. The middle marker (MM) is intended to inform the crew about impending the visual phase of landing.

OSP operation control as well as radio exchange is performed from the control tower (CTP) or command-the-starting (CSP) positions.

The crew of an airplane approaching the airdrome establishes a two-way radio contact with an air traffic controller and having received a clearance to land states its position, flight altitude, remaining fuel, etc. The controller informs the crew of landing conditions (heading, flight altitude, weather conditions in the terminal area, etc.).

To guide the aircraft to the terminal area, the crew uses the ADF and determines a LOM relative bearing. The correct capture of the landing heading is controlled by switching over the ADF from the LOM to the LMM. If ADF readings do not change, there are no deviations from the final course.

The aircraft starts descending within the specified distance to the runway, and the LRA controls its vertical height. When the aircraft passes over the marker beacons, their signals are registered by the airborne marker receiver. The height should be near 200 m when passing over the OM and near 60 m when passing over the MM. The further descent and landing is accomplished visually using the ground-based lighting equipment of the landing system.

The DF is included as part of the equipment to increase the reliability of the landing system. It allows the crew to control the aircraft attitude relative to the airdrome and calculate the approach if the ADF fails. In this case, the crew controls the aircraft using ATC instructions.

The advantage of the OSP is the simplicity of the ground and onboard equipment. The drawback of the OSP system is the impossibility of control, identification and aircraft's movement control in the external zone, in the airfield's area and at descending in complicated weather conditions. The OSP does not meet the requirements of the categorized landing (the lowest weather minimum is DH = 120 m, RVR = 1,500 m) and provides only non-precision approaches.

The OSP capacity is 15–20 aircrafts per hour. The OSP is normally used in conjunction with other landing systems as a backup. The combined landing systems are able to solve all landing problems and provide the flight safety in the terminal area for both single airplanes and aircraft groups in adverse weather conditions.

4.2.2 Radar Landing Systems

Radar landing systems (*Radiolokacionnye Sistemy Posadki (RSP)—in Russian*) provide landing of aircraft not equipped with airborne radio-technical landing equipment. In aviation, this landing management is called the ***ground-controlled approach (GCA)*** based on primary radar images. From the ground, the GCA determines the aircraft position relative to the glide path and its distance to the touchdown zone.

The GCA normally uses information from a precision approach radar (PAR) for precision approaches with vertical, glide path guidance. It assists to control the aircraft attitude in the course and descending (gliding) planes relative to the runway. If the aircraft diverts from the set values of the final course and the glide path, the controller issues on VHF the instructions for the pilots to guide them to landing. These instructions include both descent rate (glide path) and heading (course) corrections necessary to follow the correct approach path.

The GCA sometimes uses information from an airport surveillance radar (ASR), providing a non-precision surveillance radar approach with no glide path guidance.

The basic version of the radar landing system includes PAR, ATC equipment (ASR, DF), and OSP equipment to guide the aircraft to the terminal area.

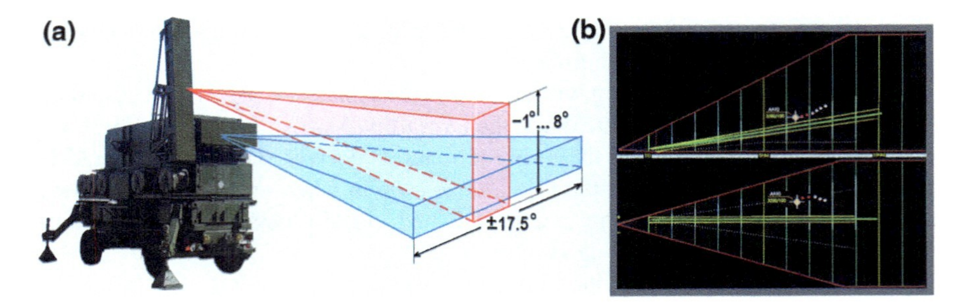

Fig. 4.4 RAP surveillance zones and indication

The ASR provides the continuous monitoring of general air situation in the terminal area within a radius of 70–80 km, approach traffic management (within a range of 250 km), and holding flights management.

The DF is included as part of the equipment to increase the reliability of the landing system and to provide the additional control of the aircraft position during approach.

The PAR is the primary equipment of the radar landing system and is usually installed next to the runway at the distance of 130–150 m from its centerline. The PAR operates in the SHF band and provides a simultaneous determination of azimuth, elevation angle, and range of the aircraft relative to the runway.

The azimuth is calculated relative to the runway centerline, elevation angle relative to the skyline, and the range relative to the optimal touchdown zone.

The PAR has a localizer and glide path antennas, each forming a pattern beam which is narrow in the plane of the measured parameter. The beams provide the surveillance of the desired airspace segments (segments up to $\pm$ 17.5° are for the localizer antenna and segments from -1 to 8° are for the glide path antenna) (Fig. 4.4a). The beam width angle of the localizer antenna in the horizontal plane is equal to 1°, and the beam width angle of the glide path antenna in the vertical plane is equal to 0.5°.

The range accuracy of modern PARs is about 15 m. The range coverage is 40–60 km. To increase the coverage range and the coordinate measuring accuracy, there is an operation mode with an airborne transponder which receives PAR signals and emits reply signals. The transponder signal can also contain additional information about the flight height, tail number, and remaining fuel.

The PAR indication system (Fig. 4.4b) has two indicators for the glide path and for the heading where the descending line and course line are shown. The upper portion of the display indicates elevation and the lower portion azimuth.

The advantages of the RSP are:

opportunity for any aircraft to land;
system mobility and a small-sized antenna system which is important for military air traffic control facilities;
the system is simple, direct and works well, even with previously untrained pilots.

The disadvantages of the system are:

It requires close communication between ground-based air traffic controllers and pilots in approaching aircraft. As a rule, only one pilot is guided at a time, thus reducing capacity (15–20 airplanes per hour).
There is no airborne indication of position relative to the glide path.
The performance of ground staff is complex.
The ground equipment is complex.

The RSP can provide only CAT I precision approaches.

4.2.3 Radio Beacon Landing Systems

Radio beacon landing systems fully meet the requirements placed on the landing systems. They continuously supply necessary information on board the aircraft and provide its indication on the corresponding displays. The information includes:
aircraft position relative to the landing course plane;

aircraft position relative to the descent (glide slope) plane;
distance to the runway threshold by means of markers (in two or three fixed points of the glide path) and to the touchdown point by means of DME (if available).

The radio beacon landing system includes radio beacons and ATC equipment as well as radio-technical and lighting facilities belonging to the OSP.
The following systems are distinguished depending on the frequency range used:

Microwave Landing System (MLS):

angle measurement channel 5,030–5,091 MHz, distance measurement channel (DME/P) 960–1,215 MHz;

UHF Landing System (*RSBN's landing channel, Russia*):

heading channel 905.1–932.4 MHz, glide path channel 939.6–966 MHz, distance measurement channel in air-to-ground direction 772–808 MHz, in ground-to-air direction 939.6–966.9 MHz;

VHF Instrument Landing System (ILS):

heading channel 108–112 MHz, glide path channel 329–335 MHz, marker channel 75 MHz, distance measurement channel (provided the system includes DME transponder) 960–1,215 MHz.
The MLS is designed to provide the flight of an aircraft along any curved four-dimensional approach trajectory in the space segment covering a sector of ±40° relative to the runway centerline in horizontal plane and a sector of 1–15° in vertical plane.

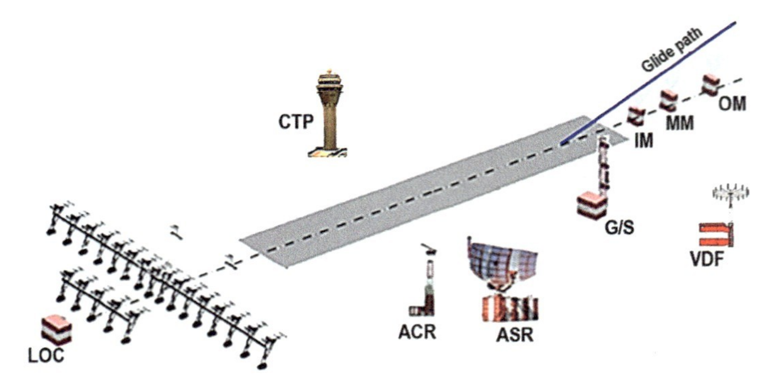

Fig. 4.5 Content and placement of the ground equipment of the instrumental landing system

The MLS can provide guidance of aircraft to the landing trajectory, takeoff, go-around and automatically transmit necessary technical and meteorological information on board the aircraft.

The ground equipment of MLS includes azimuth, distance, and elevation radio beacons. The expanded version also includes a beacon of back azimuth and an elevation alignment radio beacon.

The MLS leaves some freedom of choosing an approach trajectory. This enables to implement highly efficient approach and landing patterns, thus increasing airport capacity.

The radio beacons of UHF and VHF landing systems set a single three-dimensional approach trajectory. The equipment of these systems enables the aircraft to determine the deviation from the set trajectory (glide path).

The ILS determines the moments of passing two or three points on the glide path situated at a certain distance relative to the runway threshold. The UHF landing system (RSBN) provides continuous measurement of distance to the runway threshold.

The ILS radio beacon equipment (Fig. 4.5) includes localizer (LOC), glide slope (G/S), and marker beacons. Signals from the radio beacons are received by special airborne landing equipment (either localizer, glide slope, marker receivers, or integrated navigation and landing equipment).

ATC equipment comprises:

ASR—for monitoring the air situation in the range of 200–250 km and for controlling the approach traffic during their approach, holding, and landing;

Airfield control radar (ACR)—for management of aircraft on the airfield. It provides control of aircraft moving along the taxiways and runways;

DF—for identification of single airplanes and aircraft groups arriving over the airdrome and for determination of their azimuth at the distance of 100–150 km by onboard VHF radio station direction finding;

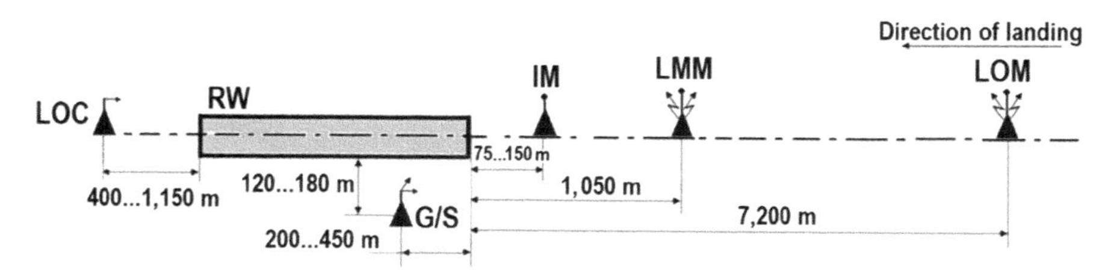

Fig. 4.6 Placement of the ILS equipment according to the ICAO standards

DME transponder—to form the reply signals for airborne interrogators. It helps to measure on board the aircraft the slant range distance to the touchdown point;

Airfield radio communication instrumentation—HF and VHF radio stations each of which services flights in the provided airspace at a specific frequency.

When using the radio beacon landing system, the aircraft heading to the terminal area and approach are performed by means of OSP (NDB) as well as ASR and DME. These aids provide entry into the LOC coverage, flight through the LOC area, and entry into the G/S coverage.

Landing accuracy in accordance with the LOC and G/S signals is checked using indicators in the cockpit. An aircraft can descend along the glide slope line in conditions of no visibility up to the height of 15–30 m. The further descent of aircrafts and their landing are accomplished on the basis of visual acquisition and airdrome lighting equipment.

The placement of the radio beacons at the airdrome according to the ICAO standards is shown in Fig. 4.6.

The LOC antenna system is installed on the centerline of the runway on the opposite side at a distance at most 1,150 m providing a safe altitude of passing over the obstacle. The distance between the G/S antenna system and the threshold must be such that the height of the reference point of the landing trajectory (a point of the glide path or extended glide path situated on the runway centerline over its threshold) is equal to 15^{+3}_{-0} m. It depends on the nominal glide slope angle, surface slopes, and other factors.

If opposite directions of approach of the runway are served by different landing systems, operation of only one of them is provided, and the other is blocked.

Marker beacons are located along an extension of the runway centerline and identify particular locations on the approach. Ordinarily, two beacons are included as part of the ILS: an OM at the final approach fix (typically 7,200 m from the approach end of the runway) and a MM located 1,050 ± 150 m from the runway threshold. The MM is located so as to note impending visual acquisition of the runway in conditions of minimum visibility for CAT I ILS approaches. An inner marker (IM), located approximately 75–150 m from the threshold, is normally associated with CAT II and CAT III ILS approaches and designates the moment of passing DH.

The capacity of the ILS is approximately 30 aircraft per hour. It can provide CAT I, II, and III A, B approaches.

The advantages of the radio beacon landing system are:

linearity of the glide path and capability to adjust its slope;
continuity and simplicity of determining the aircraft position relative to the glide path;
high accuracy which provides a safe descent of an aircraft to the height of 15–30 m;
rather high capacity.

The disadvantages of the radio beacon landing system are:

big dimensions of antenna systems which cause complex deployment of ground equipment and adjustment of LOC and G/S radiation sectors;
need for special airborne equipment to receive and process signals of landing beacons;
significant influence of earth surface and terrain characteristics around the radio beacons on the quality of operation which causes the use of expensive control systems and special provision of airdrome;
high cost of construction and installation works and operation;
small size of a sector which provides proportional dependence of an information signal on the aircraft displacement relative to the landing trajectory, thus reducing ability of lineup correction and limiting the airport capacity;
inadequacy of the facilities to the performances of promising aircraft as they provide only one straight landing trajectory which is fixed relative to the ground;
a small number of frequency channels (40 in the ILS).

The last three mentioned drawbacks do not concern the MLS system which is the most perfect of the radio beacon landing systems in terms of their performances. However, due to the prospects for the development of GNSS-based landing technologies, the MLS have not found wide application.

The current worldwide standard system for precision approach and landing is the ILS. Nowadays, the RSBN-based landing system (Russia) is used only in military aviation.

4.3 Instrument Landing System

4.3.1 Principles of Construction and Operation

The ILS is constructed on the basis of ground-based radio beacons: a localizer (LOC), a glide slope (G/S) beacon, and a marker radio beacon. The LOC transmits in the VHF band (40 frequency channels within the band of 108–112 MHz). The G/S transmitter operates in the UHF band (40 frequency channels within the band of 329–335 MHz) and is frequency paired with the LOC. Marker beacons transmit at 75 MHz in the VHF band.

To identify the radio beacons of a particular ILS, the LOC transmits an identification signal containing two or three Morse code letters the first of which is I. The identification signal is sent in groups with the frequency of 7 groups per minute.

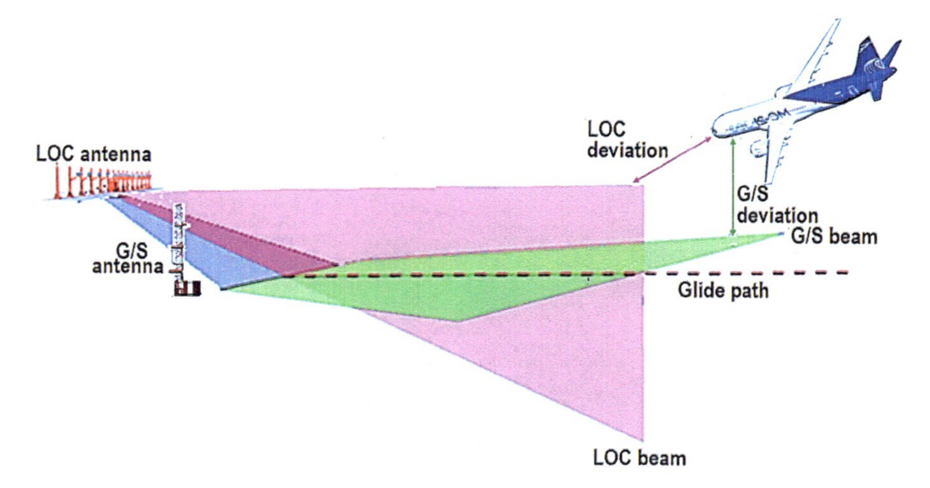

Fig. 4.7 Principle of glide path formation

The antenna systems of LOC и G/S radio beacons form radiation fields defining a course plane (a vertical plane passing through the runway centerline) and a descending (gliding) plane (a plane sloped to the runway at an angle of approximately 3° and passing through the optimal touchdown point on the runway). Intersection of these planes determines the descending line (glide path) (Fig. 4.7).

The main methods of determining the direction by means of glide slope beacons and localizers are an ***equisignal*** and ***carrier and side band/side band only*** (***CSB/SBO***) (*zero-referenced method—called in Russia*) methods.

Equisignal directions (ESD) are created by two overlapping lobes of the antenna patterns formed by LOC or G/S antennas. The LOC antenna creates horizontal ESD coinciding with the course plane. The G/S antenna creates vertical ESD coinciding with the descending (gliding) plane.

To determine the side of deviation from the ESD, the airborne equipment uses amplitude modulation of radiation lobes at frequencies of 90 and 150 Hz. The frequency of modulation is 90 Hz for the LOC left-hand lobe and 150 Hz for the LOC right-hand lobe. The G/S upper lobe is modulated with 90 Hz, lower lobe with 150 Hz.

To determine the side and value of deviation from ESD, the airborne equipment compares depths of modulation (DoM) of signals received via lobes.

The antennas of the marker radio beacons have rather narrow antenna patterns which are oriented vertically (Fig. 4.8). If the aircraft is within the scope of antenna pattern, the signal of marker radio beacon is received and visual and aural signals are triggered. This enables to define the type of the marker beacon (OM—blue cockpit light, MM—amber light, IM—white light) and thus the distance to the threshold at a given time.

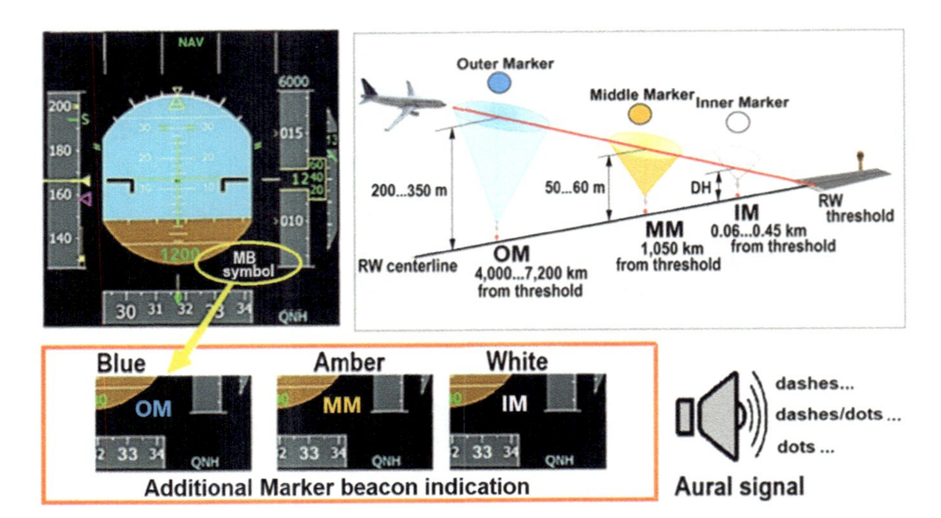

Fig. 4.8 Indication of passing over the marker beacons

The signals of marker radio beacons are amplitude-modulated with Morse code (dots and/or dashes) and differ by modulation frequencies (OM—400 Hz, MM—1,300 Hz and IM—3,000 Hz). The OM is identified by 2 dashes/s, the MM by alternate dots and dashes 2/s, the IM by 6 dots/s.

The ILS can also determine the range to the touchdown point using a ground-based transponder beacon installed next to the G/S. The antenna of DME transponder can be mounted on the same mast as the G/S antennas.

The airborne navigation and landing equipment receives and processes signals of ground-based ILS radio beacons. Its output signals contain information about the side and value of deviation of the aircraft from the course surface and from the glide path. These signals are sent to the course deviation indicator (CDI) (Fig. 4.9a) or horizontal situation indicator (HSI) (Fig. 4.9b) or to the multifunction ND (Fig. 4.9c) and PFD (Fig. 4.9d) of modern aircraft and are used by the crew or FMS to maintain the glide path accurately.

If the aircraft approaches and descends along the given glide path line correctly, the vertical and horizontal bars intersect in the center of the CDI (HIS) scale. On the ND, the index indicating deviation from the course surface coincides with an imaginary course line, and on the PFD, the indexes indicating deviation from the course and descending (gliding) surfaces are located in the center of the corresponding scale.

Figure 4.10 is a fragment of an airdrome chart with installed ILS equipment. The ILS is depicted as a sector of the so-called localizer beam. The ILS data box contains information about the landing heading (117°), LOC operating frequency (111.3 MHz), call signs (three letters of Morse code—ICN).

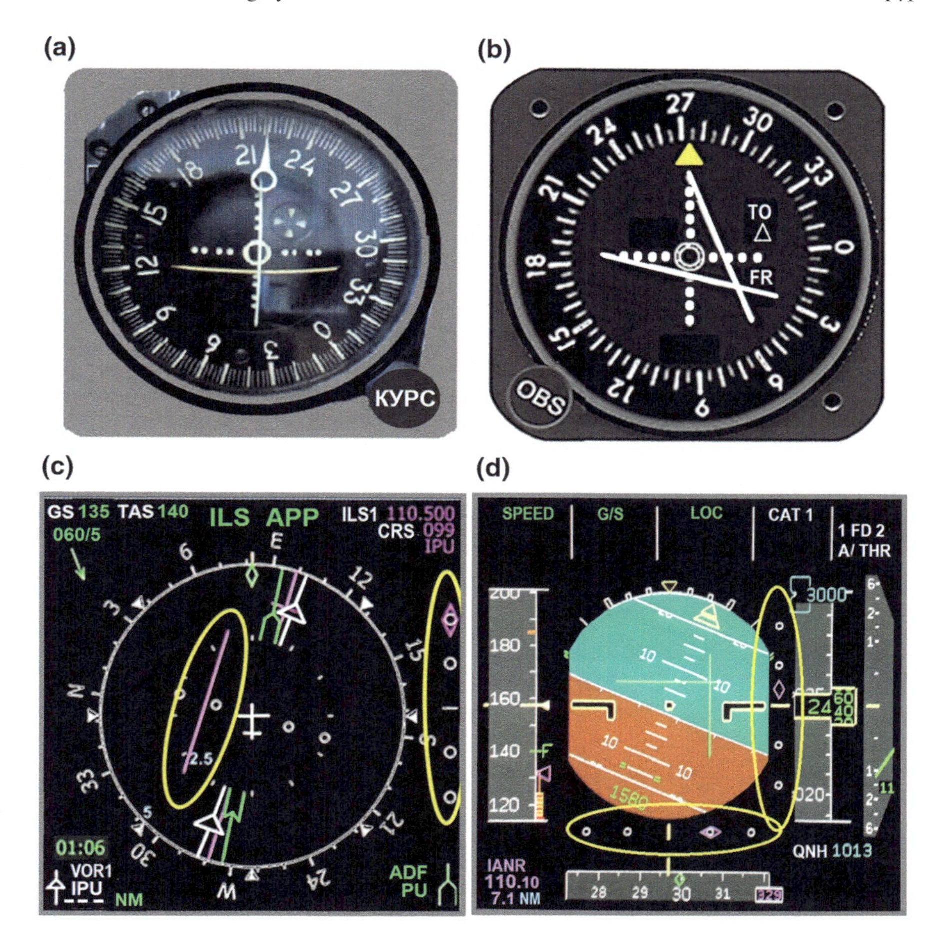

Fig. 4.9 Indicating devices of airborne landing equipment

4.3.2 Equisignal Radio Beacons

Simplified block diagrams of LOC and G/S implementing the equisignal method can be presented in a general block diagram shown in Fig. 4.11.

VHF oscillator (VHFO) forms carrier oscillations. The oscillations are amplitude-modulated in amplitude modulators AM1 и AM2 with low-frequency 150 Hz oscillations which are formed by a low-frequency oscillator LFO1 and with oscillations 90 Hz which are formed by oscillator LFO2. Amplitude-modulated signals are supplied to antennas A1 and A2 having overlapping antenna patterns $F_1(\Theta)$ and $F_2(\Theta)$.

The line of intersecting the antenna patterns $F_1(\Theta)$ and $F_2(\Theta)$ forms an equisignal direction (ESD) coinciding with the course (glide slope) line.

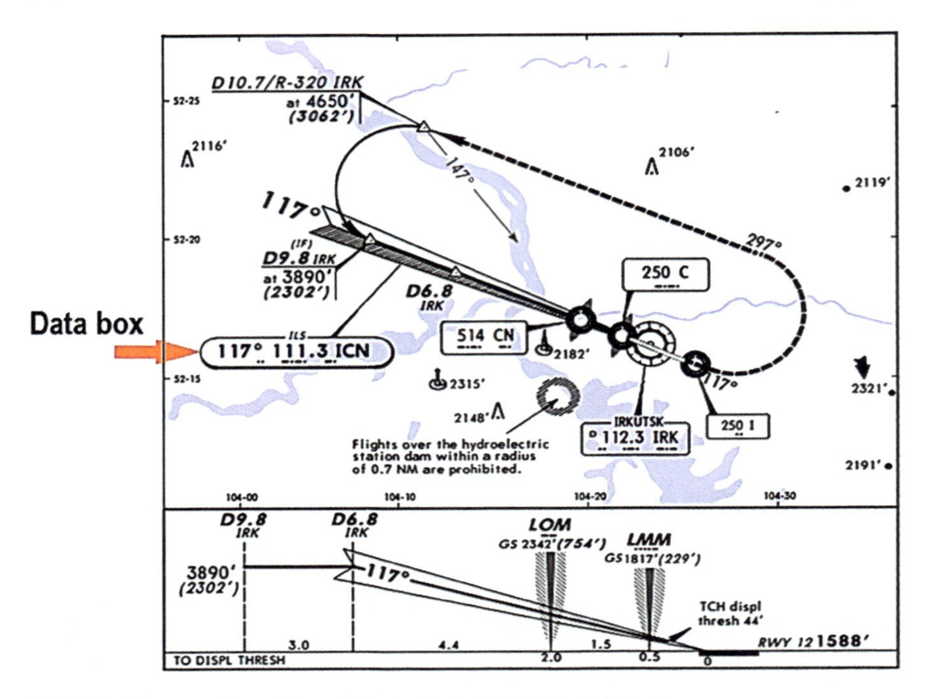

Fig. 4.10 Fragment of an airdrome chart with installed ILS equipment

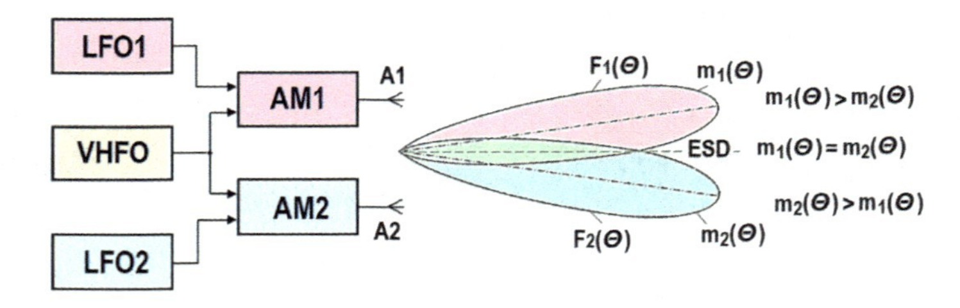

Fig. 4.11 Block diagram of an equisignal radio beacon

The antennas form two radiation fields with modulation frequencies Ω_1 and Ω_2:

$$e_1 = E_m F_1(\Theta)(1 + m \cos \Omega_1 t) \cos \omega t,$$
$$e_2 = E_m F_2(\Theta)(1 + m \cos \Omega_2 t) \cos \omega t,$$

where E_m are field amplitudes at antenna outputs which are assumed to be equal; m are indexes of amplitude modulation of emitted signals which are also assumed to be equal.

In space, the amplitudes of the fields depend on the position of a receiver relative to antenna patterns $F_1(\Theta)$ and $F_2(\Theta)$.

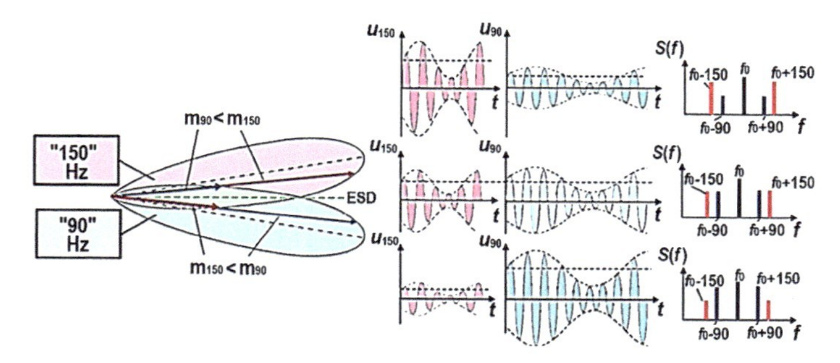

Fig. 4.12 Influence of the receiver position on characteristics of the resultant field in the reception point

When combining the fields produced by antennas A1 and A2, a resultant radiation field amplitude-modulated with frequencies of 150 and 90 Hz is formed in space as:

$$\begin{aligned} e_{\Sigma} &= e_1 + e_1 \\ &= E_m[F_1(\Theta) + F_2(\Theta)]\{1 + M_1(\Theta)\cos\Omega_1 t + M_2(\Theta)\cos\Omega_2 t\}\cos\omega t, \end{aligned}$$

where variables:

$$\begin{aligned} M_1(\Theta) &= mF_1(\Theta)/[F_1(\Theta) + F_2(\Theta)] \\ M_2(\Theta) &= mF_2(\Theta)/[F_1(\Theta) + F_2(\Theta)], \end{aligned}$$

are depths of space modulation (DoSM).

The airborne equipment extracts oscillation modulation envelopes of frequencies Ω_1 and Ω_2 and determines the difference in modulation envelope amplitudes which is proportional to the difference in depth of modulation (DDM).

The value and sign of DDM $= M_1(\Theta) - M_2(\Theta)$ gives an indication of aircraft position relative to the course (glide slope) planes.

The spectrum of the resultant radiation field of an equisignal radio beacon is shown in Fig. 4.12. The amplitudes of spectral components of the resultant field depend on the depths of space modulation DoSM which in their turn depend on deviation of the aircraft from the equisignal direction. The spectral components are in-phased at all side frequencies and at the carrier.

If an aircraft flies in the equisignal direction ESD, the amplitudes of spectral components of 90 and 150 Hz modulation frequencies are equal. If the aircraft deviates left or right (up or down) from the equisignal direction ESD, the amplitude of the component of 90 Hz modulation frequency increases or decreases and the amplitude of the component of 150 Hz modulation frequency, on the contrary, decreases or increases.

Formation of a G/S radiation field is greatly influenced by the terrain near its antenna. In the UHF/VHF band, the earth surface is like a conductor and effectively

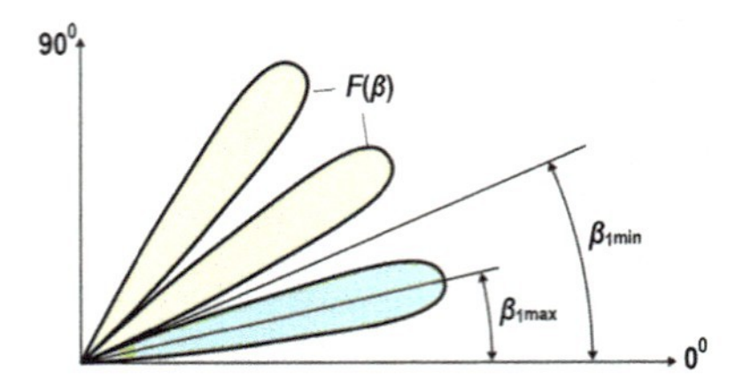

Fig. 4.13 Multilobe antenna pattern in vertical plane

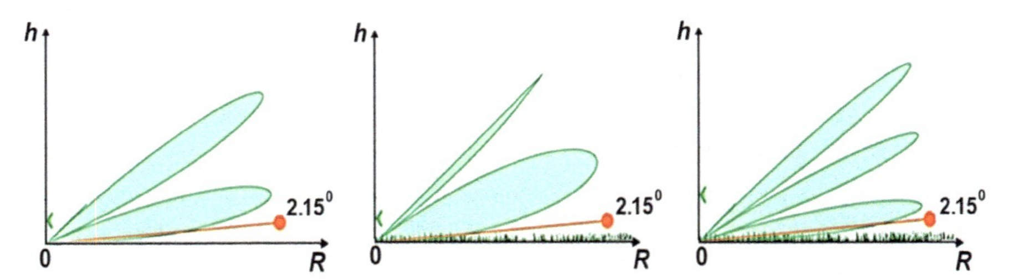

Fig. 4.14 Influence of the height of antenna mounting and the character of underlying terrain in the form of antenna pattern

reflects incident radiation into the upper hemisphere. Therefore, the resultant G/S antenna pattern is formed by interfering the direct and earth-reflected signals which are combined with different phase shifts in different points of space.

Due to the interference of the direct and earth-reflected signals, the antenna pattern acquires a multilobe character in vertical plane. The maxima (peaks) of the resultant antenna pattern are at angles $\beta_{\max}$ where the signals are in-phased, while its dips are at angles $\beta_{\min}$ where the signals are anti-phased (Fig. 4.13).

Directions of antenna pattern maxima and minima can be defined by equations:

$$\sin\beta_{\min} = \frac{k\lambda}{2h};\ \sin\beta_{\max} = \frac{(2k+1)\lambda}{4h};\ k = 0, 1, 2, \ldots,$$

where h is a height of antenna mounting above the ground, and λ is a wavelength of radiated oscillations.

Tilt angles of antenna pattern maxima and minima relative to the ground are determined by relationship between the height of antenna mounting and the wavelength. They are also influenced by state of the earth's surface in vicinity of the antenna which causes variations of the ground reflection coefficient (Fig. 4.14). If the height of antenna mounting increases, the number of vertical lobes in the antenna pattern rises, while the values of tilt angles of the first minimum and maximum decrease.

To implement the equisignal method of specifying the glide path, two overlapping antenna patterns are needed. They are formed by two antennas installed at different heights h_1 and h_2. The upper antenna has more lobes than the lower one.

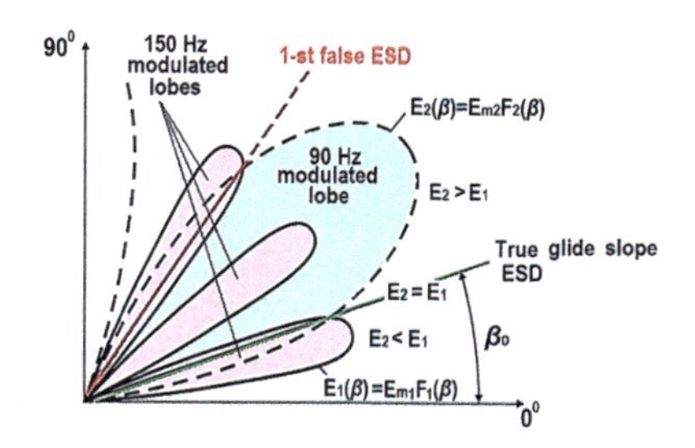

Fig. 4.15 Formation of false glide paths

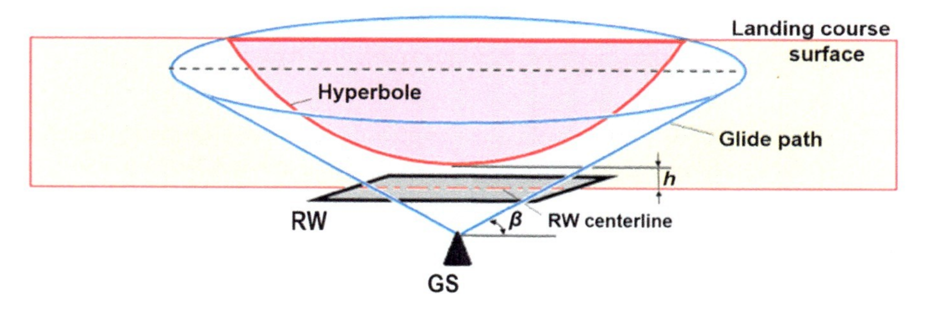

Fig. 4.16 Concerning the nonlinearity of the glide path

Since both antennas have multilobe antenna patterns, several equisignal directions are formed in the vertical plane (Fig. 4.15). The desired tilt angle of the first equisignal direction (true glide slope) and rather big (more than 15°) tilt angle of the first false glide slope can be provided by adjusting the height of antenna mounting and radiation power supplied.

It is necessary to note one more peculiarity of the formed ILS glide path. In Figs. 4.1 and 4.7, the glide path is shown as a straight line which is formed by intersecting the course and descending (glide slope) planes and touches the runway surface. In fact, the intersection of the planes formed by the equisignal directions of LOC and G/S antennas makes up a hyperbole (Fig. 4.16). This is caused by the displacement of G/S antenna from the runway centerline by 120–180 m.

Indeed, if the G/S antenna pattern is rotated in the horizontal plane, its ESD will describe a surface in the form of a cone with the center in the point of beacon placement. The ESD specified by LOC is in the vertical plane passing through the runway centerline. The line of intersection of the cone surface with the vertical plane displaced (shifted) about the cone axis is a hyperbole situated at height h above the ground. In this case, $h = L \cdot tg\ \beta$ where L is a distance between the G/S antenna and the runway centerline. With typical values $\beta = 2°\ 40'$ and $L = 150$ m we will receive $h = 7$ m.

Due to deployment of LOC and G/S antenna systems and setting of radiation sectors and power, it is possible to achieve the alignment of the glide path on final approach and its maximum proximity (up to 1.5 m) to the runway surface.

If an aircraft is on the glide path and there are slight deviations from the equisignal directions, low-level signals enter the receiver inputs of the equisignal radio beacons. The equisignal radio beacons are, therefore, vulnerable to the secondary emissions from roughness of relief and objects situated in the radiation sector as well as to other factors causing the glide path bend. Thus, the equisignal radio beacons meet the requirements of CAT I operation (not better).

4.3.3 "Zero-Referenced" (CSB/SBO) Radio Beacons

"Zero-referenced" (CBS/SBO) radio beacons provide a course (glide path) line with higher stability. They work using an advanced equisignal method.

In contrast to an equisignal beacon, a "zero-referenced" radio beacon contains balanced modulators BM1 and BM2, a sum-and-difference bridge, and an antenna distribution network (ADN). The BM multiplies instantaneous values of oscillations of high frequency and either low frequency Ω_1 or Ω_2 (Fig. 4.17). The peculiarity of BM oscillations is the absence of HF (carrier) oscillations in their spectrum. The spectrum of BM oscillations includes only oscillations of side modulation frequencies (side band only (SBO)) $\omega \pm \Omega_1$ and $\omega \pm \Omega_2$ since:

$$\cos \Omega t \cos \omega t = [\cos(\omega - \Omega)t + \cos(\omega + \Omega)t]/2.$$

Signals from BM outputs enter the sum-and-difference bridge whose outputs (an adder and a contractor) supply the signals to the antennas.

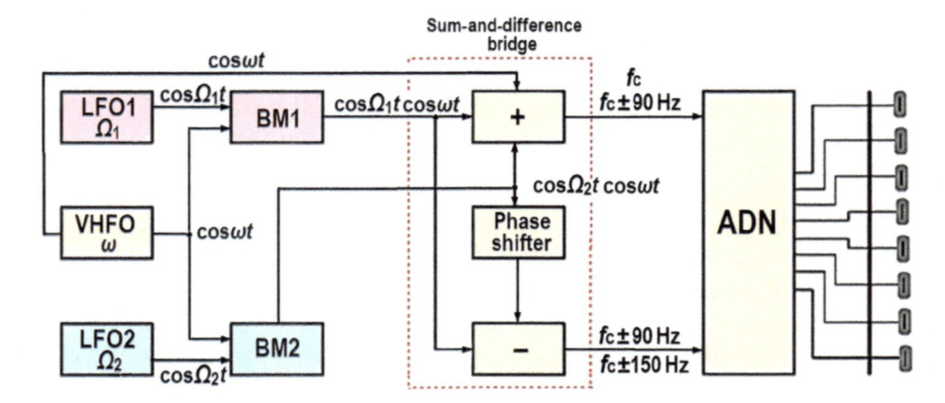

Fig. 4.17 Simplified block diagram of a "zero-referenced" radio beacon

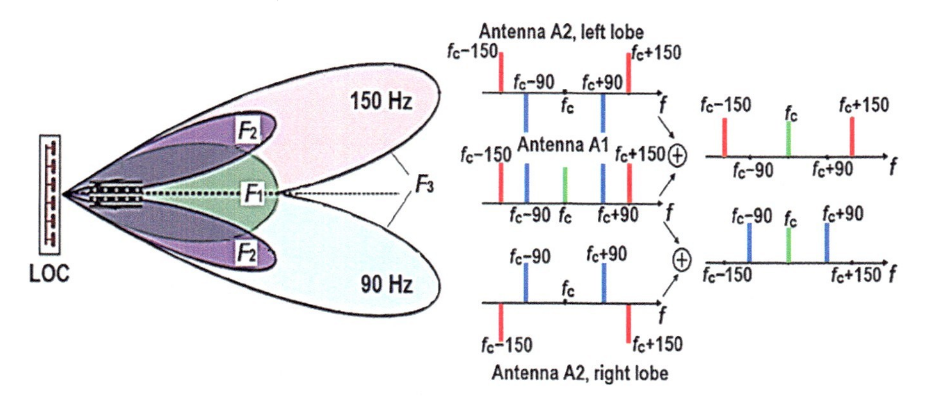

Fig. 4.18 LOC sum antenna radiation pattern

Multi-element "linear-series" antennas are used in LOC. The antenna consists of near-omnidirectional antennas (e.g. wave-channel or loop antennas) aligned perpendicular to the runway centerline at regular intervals from each other. The center of a linear series is on extension of the runway centerline. The emitters of antenna are fed at specified amplitudes and phases to ensure a desired antenna radiation pattern.

Via the ADN, all the antenna emitters receive equiphase signals from the adder. This creates a radiation field of antenna A1 as:

$$e_1 = E_{m1} F_1(\Theta)(1 + m \cos \Omega_1 t + m \cos \Omega_2 t) \cos \omega t,$$

which is amplitude-modulated with oscillations of frequencies Ω_1 and Ω_2. It is characterized by a single-lobe antenna pattern $F_1(\Theta)$ whose maximum coincides with the runway centerline (Fig. 4.18).

The spectrum of A1 radiation field contains a carrier f_c and side components at frequencies $f_c \pm 150$ Hz and $f_c \pm 90$ Hz (carrier and side band (CSB)).

Signals from the contractor (side frequencies) enter the antenna emitters via the ADN so that the current phases of dipoles on the right and left sides of the centerline are opposite. The side components with modulation frequencies of 90 and 150 Hz are anti-phased relative to each other as well.

With such phase relations of supplying oscillations, the antenna system is equivalent to two symmetrically spaced anti-phased dipoles which form antenna A2. The antenna A2 creates a twin-lobe antenna radiation pattern $F_2(\Theta)$ whose minimum coincides with the runway centerline (Fig. 4.18).

The radiation field of antenna A2 contains only side frequencies of modulation:

$$e_2 = E_{m2} F_2(\Theta)(\cos \Omega_1 t \ - \cos \Omega_2 t) \cos \omega t$$

As a result of combining the radiation fields of antennas A1 and A2, a resultant radiation field is formed and its depths of space amplitude modulation with frequencies Ω_1 and Ω_2 depend on angle Θ as:

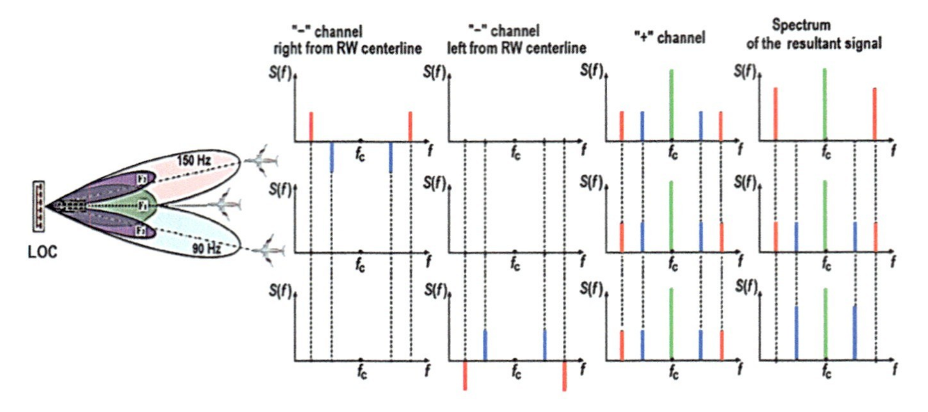

Fig. 4.19 Influence of the receiver position on the characteristics of the resultant field in the reception point

$$e = e_1 + e_2 = E_{m1} F_1(\Theta)[1 + M_1(\Theta)\cos\Omega_1 t + M_2(\Theta)\cos\Omega_2 t]\cos\omega t,$$

where the depths of space modulation are:

$$M_1(\Theta) = m + aF_2(\Theta)/F_1(\Theta)$$
$$M_2(\Theta) = m - aF_2(\Theta)/F_1(\Theta),$$

where $a = E_{m2}/E_{m1}$, m is a DoM of radiated oscillations.

This resultant field corresponds to the field formed in space by an antenna with a twin-lobe antenna radiation pattern $F_3(\Theta)$ whose equisignal direction coincides with the course plane (Fig. 4.18).

The depth of space amplitude modulation of the field with frequencies of 90 and 150 Hz changes within the radiation sector of the LOC antenna system. The depth of 90 Hz modulation of the carrier prevails to the left from the course plane, while the depth of 150 Hz modulation prevails to the right (Fig. 4.19).

On the course plane, the depths of 90 and 150 Hz modulation are equal; therefore, DDM = 0. If the aircraft moves away from the course plane, the DDM increases. Thus, the value of DDM is an indication of deviation rate in the horizontal plane, while the predominance of the depths of modulation of a specific frequency is an indication of deviation side.

A "zero-referenced" G/S has significant advantages in comparison with the equisignal one. It provides separate adjustment of the steepness of glide slope angle and area. The steepness can be smoothly adjusted by changing the level of side frequencies in the upper antenna. The change of relationship between power values of the upper and lower antennas in the equisignal G/S leads to the change of the glide path line.

"Zero-referenced" radio beacons ensure higher accuracy and stability of the course (glide path) line, thus providing CAT II and sometimes CAT III approaches. Besides, they have good possibilities of adjustment which is important under operation conditions.

4.3.4 Twin-Channel Radio Beacons

The character of course line distortion is greatly influenced by ground features in the LOC emission sector. A re-emission (re-reflection) field created by them is combined with the LOC field and leads to course line distortion (Fig. 4.20).

The character of course line distortion depends on the re-emission pattern and intensity as well as on the location of ground features relative to the LOC antenna pattern. In different points of space, phase relations between the LOC field and re-emission field change due to variations of propagation difference of direct and re-reflection waves.

Re-emission field of side frequencies of modulation is the most dangerous as it is exactly the field that defines the course line.

To increase the accuracy of LOC and G/S, it is necessary to increase the stability of position of the course line and the glide path and to reduce the distortion of these lines. This effect is achieved by narrowing the antenna radiation patterns.

The LOC, however, must have rather wide coverage in the horizontal plane, while the G/S must provide operation at low elevation angles. To meet these requirements means to widen the antenna patterns.

In doing so, more ground features and topographic inequalities are involved in creation of reflected signals which results in increase of distortions.

The contradiction is eliminated in twin-channel radio beacons whose antenna systems form antenna radiation patterns with different widths.

The width of antenna pattern of the LOC antenna array is defined by number of its components. Figure 4.21 demonstrates how the form of LOC antenna pattern is influenced by the number of emitters N and relative distance between them d/λ (*a* –

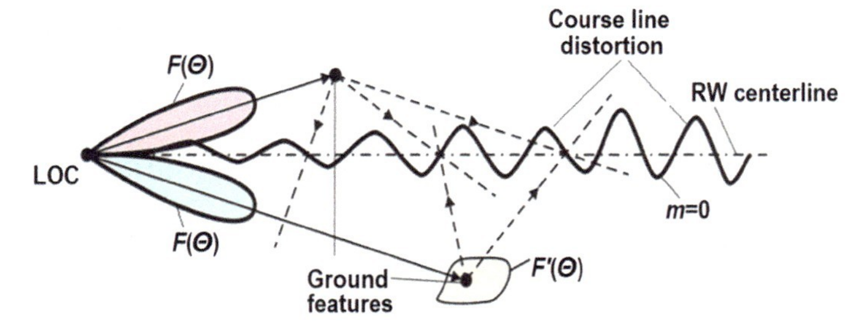

Fig. 4.20 Explanation of creating the course line distortion

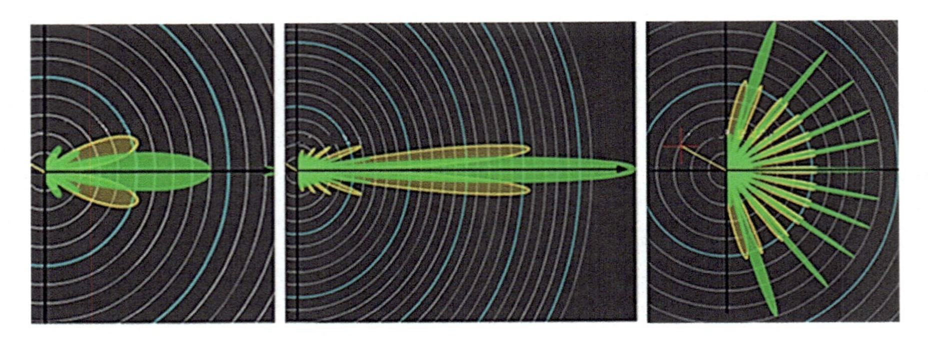

Fig. 4.21 Influence of LOC antenna parameters in the form of antenna radiation pattern

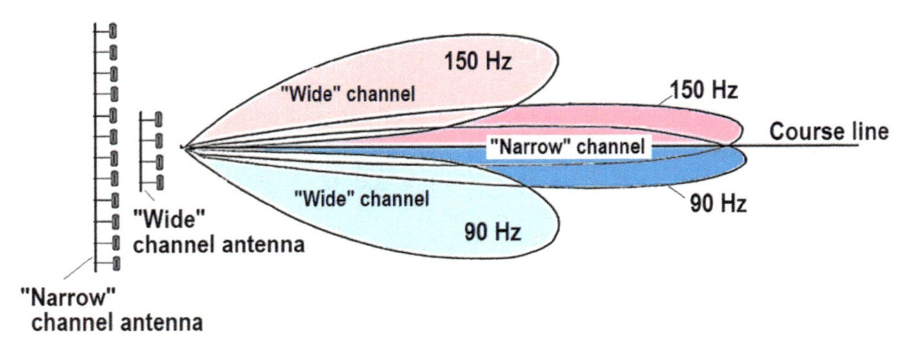

Fig. 4.22 Antenna patterns of a twin-channel LOC

$N = 6$, $d/\lambda = 0.5$; $b - N = 20$, $d/\lambda = 0.5$; $c - N = 6$, $d/\lambda = 5$). The more emitters are used, the narrower the lobes are. Violation of condition $d/\lambda \leq 0.5$ causes the multilobe character of antenna radiation pattern.

A twin-channel LOC has two channels: a "narrow" and a "wide" one, each using its own antenna system (Fig. 4.22). The antenna system of a "narrow" channel (at least 20 dipoles) forms rather narrow antenna patterns 6–12° wide in horizontal plane. The antenna system of a "wide" channel (6–8 dipoles) creates wide antenna patterns which provide the desired width of coverage (±35°).

In twin-channel radio beacons, the influence of the "wide" channel on the "narrow" one must be minimal. For this purpose, the following variants of two-channel radio beacons can be constructed:

twin-channel radio beacons with the "narrow" and "wide" channels radiating at different carrier frequencies;

radio beacons with quadrature clearance, that is their "narrow" and "wide" channels use the same frequency with modulating oscillations phase-shifted by 90° (in quadrature);

integrated radio beacons using combination of both methods.

Two-frequency radio beacons are most widely used. They are a little more complicated than radio beacons with quadrature clearance [two transmitters, stricter requirements for the stability of frequency (2×10^{-5} for a two-frequency beacon and 5×10^{-5} for quadrature clearance)], but they do not need thorough adjustment and are simple in operation. Moreover, two-frequency radio beacons suppress a re-reflected signal of a clearance channel to a greater extent.

Carrier spacing in a two-frequency LOC is from 5 to 14 kHz relative to the nominal frequency specified by ICAO for ILS. Both carriers are within the pass band of an airborne ILS receiver and the channels have hardly any effect on each other.

The "narrow" channel makes up an almost straight course line as its coverage does not include any re-reflecting ground features and topographic inequalities. This channel is used to manage aircraft at final approach with small deviations from the final course plane.

During the initial approach when considerable deviations from the final course plane are possible, the "wide" channel is used. It is vulnerable to the effects of re-reflected signals, but, for its coverage, there are no strict requirements for accuracy as large deviations from the final course plane are admissible.

In the coverage of LOC "narrow" channel, there is a radiation pattern null for the "wide" channel. It reduces the influence of "wide" channel signals upon "narrow" channel operation.

The power of "wide"-channel re-reflected signals is much lower than that of the main "narrow"-channel signal. The linear detection in the airborne receiver causes their suppression by the stronger "narrow"-channel signal. So, the "wide"-channel signals are not available on the course line and around it, thus providing high stability of radiation parameters of the LOC antenna system.

In G/S, radiation compensation at small elevation angles (up to 1.5°) is accomplished to reduce glide path distortions caused by earth re-reflections and an additional channel is used to obtain information about aircraft position in the area (Fig. 4.23).

In addition to the lower (A1) and upper (A2) antennas, a supplementary antenna *A*3 is installed three times as high as the lower antenna mounting in a twin-channel G/S.

All three G/S antennas are used to form the antenna radiation pattern of the "narrow" channel, antennas A1 and A3 participate in formation of the "wide"-channel radiation pattern.

Phases and amplitudes of currents for feeding the antennas [from the outputs of sum-and-difference bridge (SDB)] are such as to reduce the field level at small angles to the horizon. This reduces the power of signals reflected by topographic inequalities and, hence, leads to reduction of glide path distortions.

In the area above the glide path, "wide"-channel signals are suppressed by a stronger "narrow"-channel signal.

Via the "narrow" channel, the radio beacon forms two antenna radiation patterns in vertical plane. One of them is a sum pattern created by the carrier and side modulation frequencies of 90 and 150 Hz (antenna A1), and the other is a difference pattern where side modulation frequencies in adjacent lobes are anti-phased (antenna A2).

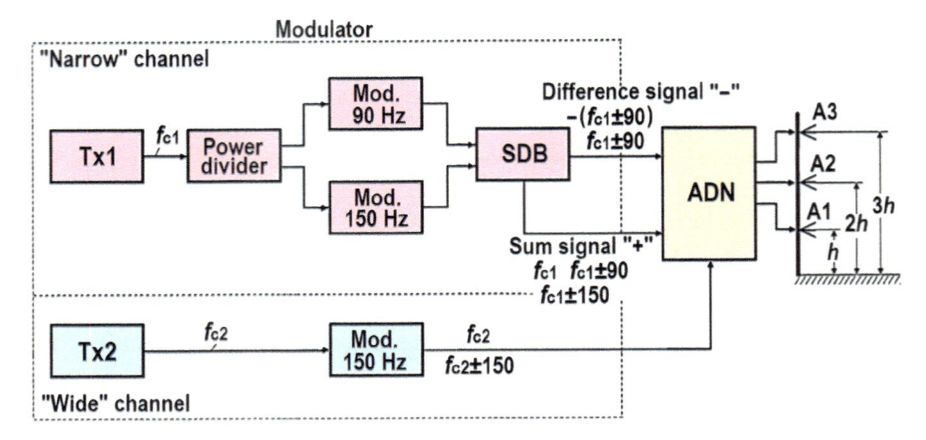

Fig. 4.23 Block diagram of a twin-channel G/S

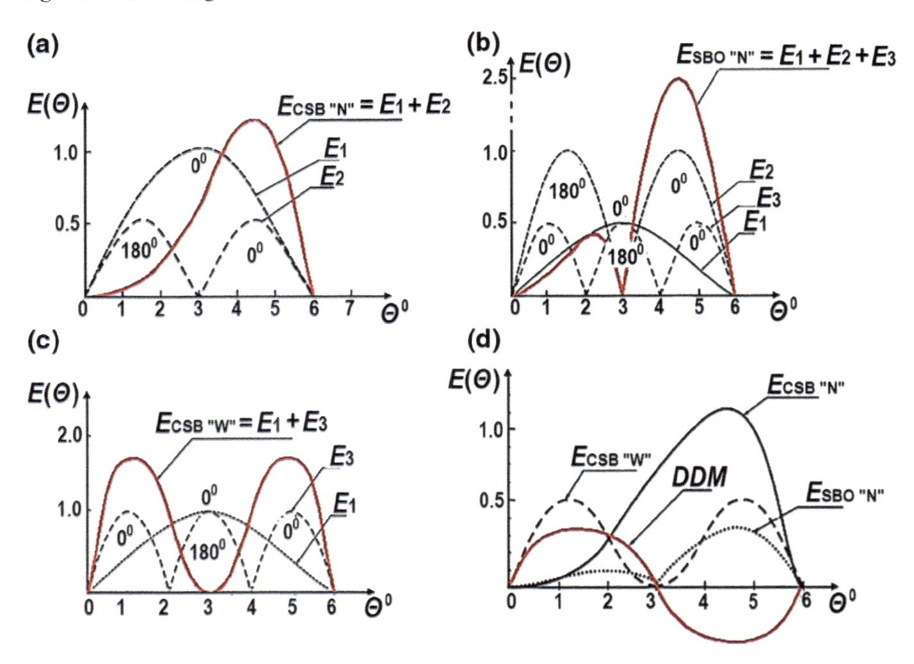

Fig. 4.24 Formation of radiation patterns of a twin-channel G/S

Combining the A1 and A2 antenna radiation fields gives a resultant antenna radiation pattern of the carrier and side modulation frequencies E_{CSB} "N" shown in Fig. 4.24a.

By signals of side modulation frequencies, antennas A1 and A3 are fed in-phase between each other and anti-phase relative to antenna A2. Antenna A2 is anti-phased

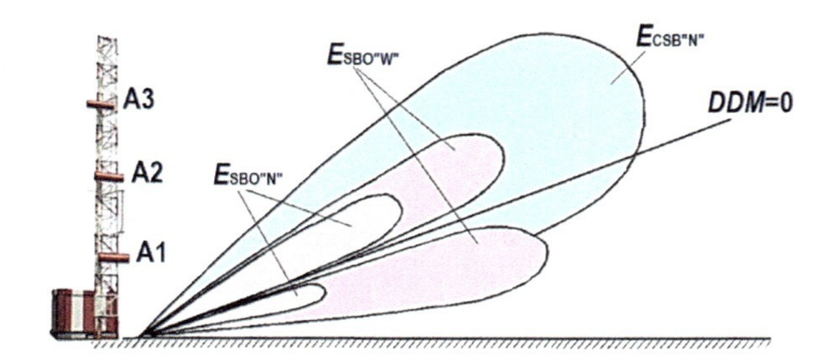

Fig. 4.25 Antenna pattern and the antenna system of a twin-channel G/S

relative to the lower antenna A1 with the purpose of suppressing the sum signal field up to angles about $1/3\beta$.

Combining the radiation fields of all three antennas gives a resultant antenna radiation pattern of the "narrow" channel E_{SBO} "*N*" shown in Fig. 4.24b.

Via the "wide" channel, antennas A1 and A3 are excited in-phase by signals of the carrier and side frequencies of the "wide" channel (modulation of the "wide"-channel carrier with 90 and 150 Hz signals with excessive 150 Hz modulation by 30%).

Combining the fields of both antennas gives a resultant antenna radiation pattern of the "wide" channel E_{CSB} "*W*" shown in Fig. 4.24c.

On the glide path line ("narrow"-channel coverage), depths of 90 and 150 Hz modulation of the carrier are equal.

In the area above the glide path, "wide"-channel signals are suppressed by much stronger "narrow"-channel signals. At small angles to the horizon, prevailing is the "wide"-channel signal which contains information only about the deviation side.

In Fig. 4.24d resultant radiation patterns of a twin-channel G/S are shown.

The form of the antenna pattern of a twin-channel G/S antenna system is shown in Fig. 4.25.

To reduce the influence of the additional channel on the main channel operation, their carriers differ by 18–20 kHz.

The heights of antenna mountings are rather high in comparison with the wavelength, so to provide desired changes of DDM in the near-field, the antennas must be displaced to the runway in such a way that the line of their placement has the form of an arc with radius R in the plane perpendicular to the runway centerline.

4.3.4.1 Processing of ILS Signals by Airborne Equipment

The ILS airborne equipment is designed to receive and process signals of radio beacons to form signals containing information about DDM. In an equisignal direction DDM is equal to zero, when diverting from it DDM increases and the sign of difference depends on the side of aircraft deviation from the equisignal direction. The signal containing information of DDM is supplied to an indicating device and FMS.

The block diagrams of localizer and glide path receivers are equal. They are also equal for equisignal and "zero-referenced" (CSB/SBO) radio beacons. A simplified block diagram of an onboard analog receiver (localizer or glide path one) is shown in Fig. 4.26.

Signals received by the antenna are amplified, converted to an intermediate frequency in the VHF module of the receiver and detected by an amplitude detector AD. Low-frequency oscillations at the AD output are separated by filters F1 and F2 tuned to the frequencies of 150 and 90 Hz. The oscillations are rectified by rectifiers R1 and R2 and enter a subtract circuit SC with a pointer-type indicator (HSI or CDI) at the output. The indicated signals are proportional to DDM, and their polarity shows the side of aircraft deviation from the equisignal direction (Fig. 4.27).

Output voltages of rectifiers are also supplied to add circuit AC whose output signal controls the flags of the indicating device ("Course readiness" and "Glide

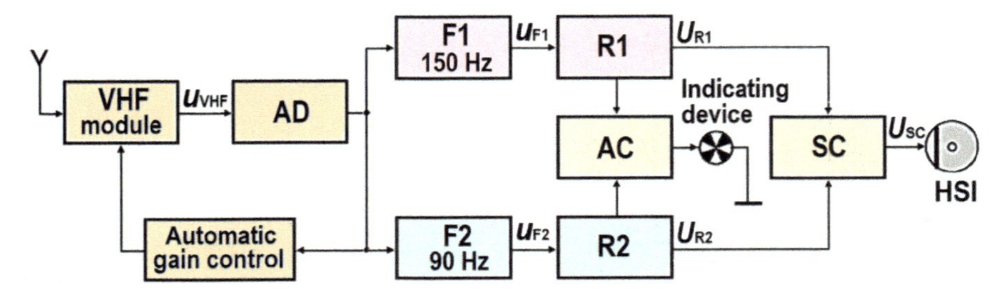

Fig. 4.26 Simplified block diagram of an onboard receiver

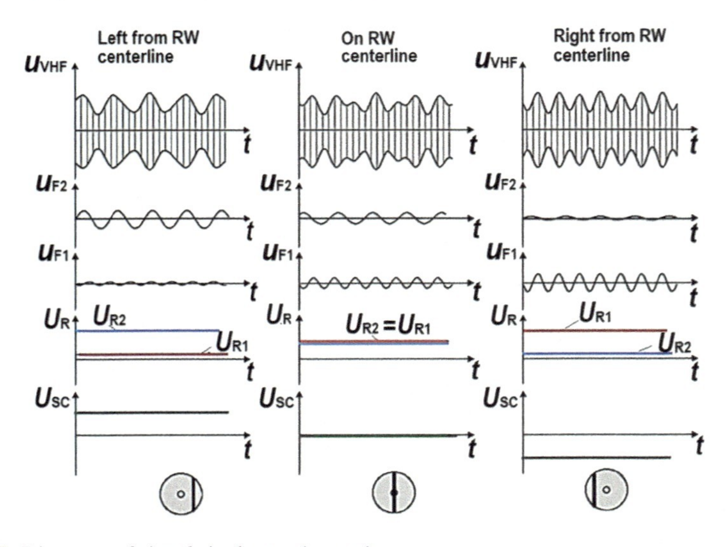

Fig. 4.27 Diagrams of signals in the receiver units

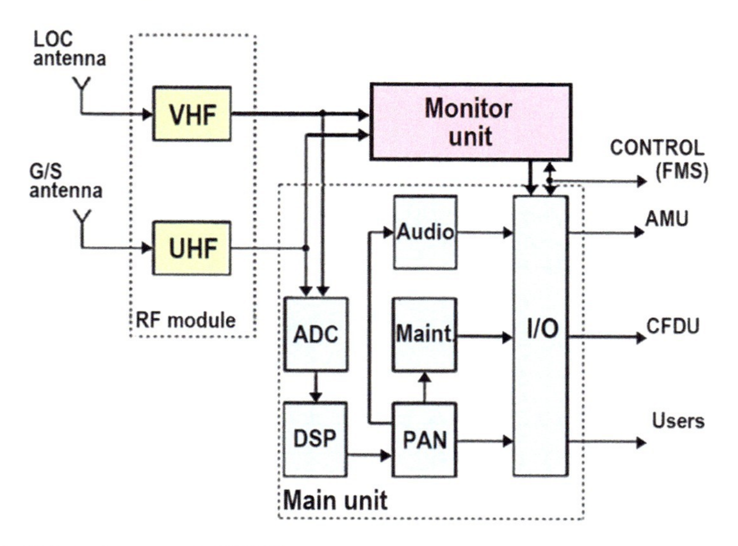

Fig. 4.28 ILS receiver as part of MMR

path readiness" signals or ILS FAILURE flag). The flags get out of sight if there are oscillations of both low frequencies 90 and 150 Hz at the rectifier outputs.

If even one of these oscillations is missing, the flag system does not come into action and the flags fit into the pilot's field of view, thus indicating that there is no signal from the localizer (glide slope) radio beacon.

On modern aircraft, the ILS receivers are included into multifunction receiving devices.

The ILS receiver (Fig. 4.28) includes three functional units: RF module, main ILS unit, and monitor ILS unit.

RF module converts (filters, mixes, amplifies, and demodulates) the signals received by VHF receiver (LOC signals) and UHF receiver (G/P signals). Analog–digital converter (ADC) transforms output analog signals from RF module into digital signals for processing by the digital signal processor (DSP).

Main unit controls ILS mode of operation, generates the audio and deviation outputs, controls the aircraft interfaces, and performs the maintenance tasks (for CFDIU interface).

The main unit is divided into five sections:

analog–digital converter (ADC);
DSP section which formats and sends to the precision approach navigator (PAN) the deviations computed with the DoM of the 90 and 150 Hz;
PAN section which processes the data from the DSP section to provide the LOC and G/S deviation information to the input/output (I/O) section. The PAN compares information from monitor unit with the information it has calculated itself. The PAN sends to the maintenance section (MAINT) the in-line test results from all the units installed in the ILS receiver;

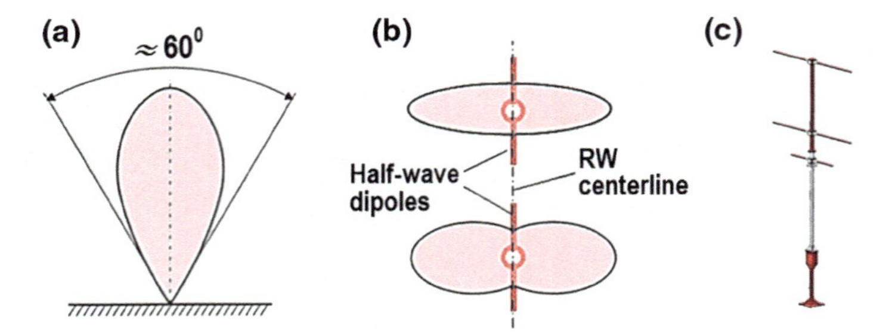

Fig. 4.29 Antenna patterns and appearance of a marker beacon antenna

MAINT which provides the interface with the centralized fault display interface unit (CFDIU);
I/O section which provides the interface with aircraft systems.

The monitor unit provides a redundant dissimilar signal processing path for the LOC and G/S signals from the ILS RF module. It also includes an ADC, a DSP, a PAN, and an I/O section (as main unit).

The monitor unit performs a validity check of the primary instrumentation processor deviation output and disables the outputs if the check fails.

The ILS receiver applies its audio output (identification signals transmitted by LOC) to the audio integrating system via audio management unit (AMU).

4.4 Marker Beacons

ILS marker channel includes marker beacons and onboard marker receivers. ILS marker beacons are used to specify points situated on the runway centerline at a certain distance from its threshold. They simplify aircraft control during approach.

The antenna of a marker beacon possesses pronounced directional properties in vertical plane and forms a cone-shaped antenna pattern (Fig. 4.29a). In horizontal plane view, the antenna pattern is lens-shaped or bone-shaped (Fig. 4.29b).

In horizontal plane, the antenna pattern has a different width along the runway centerline and in perpendicular direction. The antenna pattern width along the runway centerline must ensure that an aircraft flying with the speed of 240 km/h receives an OM signal for 12 ± 4 s and a MM signal for 6 ± 2 s. The created coverage of the marker beacon must provide operation of an onboard marker receiver when descending the aircraft along the flight path at the height of 600 ± 200 m above the OM, 300 ± 100 m above the MM and 150 ± 50 m above the IM.

The most common antenna of a marker beacon is composed of two horizontal half-wave dipoles located at a distance of quarter wavelength from the reflector (Fig. 4.29c).

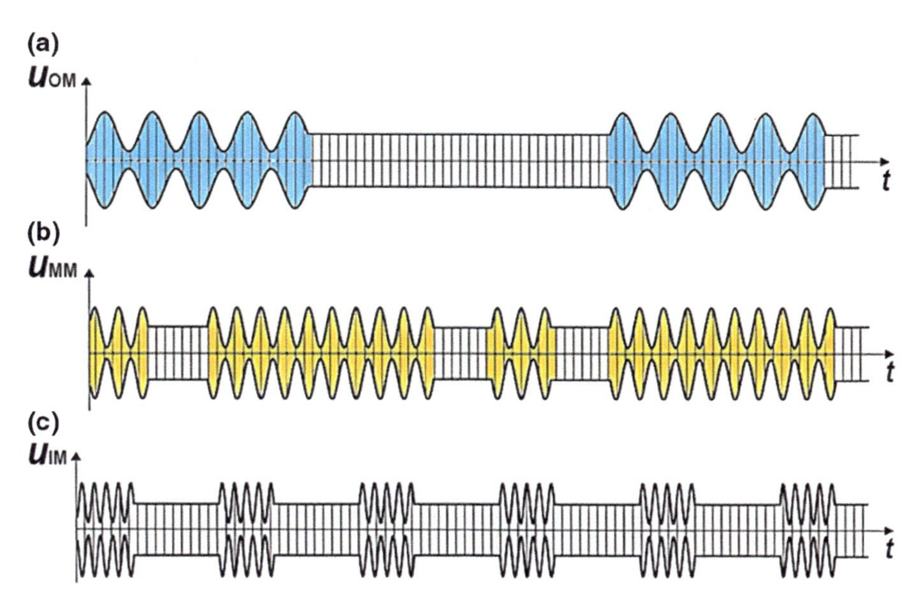

Fig. 4.30 Diagrams of marker beacon signals

The marker beacons radiate signals at a fixed frequency of 75 MHz. For identification of radio beacons, the transmitted signals include tone modulation and carrier keying by Morse code.

The marker beacons use the following combinations of modulated oscillation frequency and code of keying:

OM: modulation frequency is 400 Hz, and code of keying is 2 dashes/sec;
MM: modulation frequency is 1,300 Hz, and code of keying is alternate dots and dashes, 2 combinations/sec;
IM: modulation frequency is 3,000 Hz, and code of keying is 6 dots/sec.
Diagrams of marker beacon signals are shown in Fig. 4.30 (a—OM, b—MM, c—IM).

A simplified block diagram of a marker beacon and output waveform diagrams for the main functional units are shown in Fig. 4.31.

VHF oscillations of a high-frequency oscillator (VHFO) enter modulator MOD where they are amplitude-modulated with the amplified voltage from a low-frequency oscillator LFO (400, 1,300 or 3,000 Hz—depending on the marker beacon location). The low-frequency amplifier (LFA) operates in a switching mode. The switching (on/off) of the amplifier LFA is performed by the keyer KEY in accordance with the code of keying. The signal from the MOD output is supplied to a power amplifier (PA), and after amplification, it is sent to the antenna.

The marker radio receiver receives and processes the signal from the marker beacon on board the aircraft. The radio receiver can be installed as a self-contained unit, or it can be integrated in the airborne navigation and landing equipment.

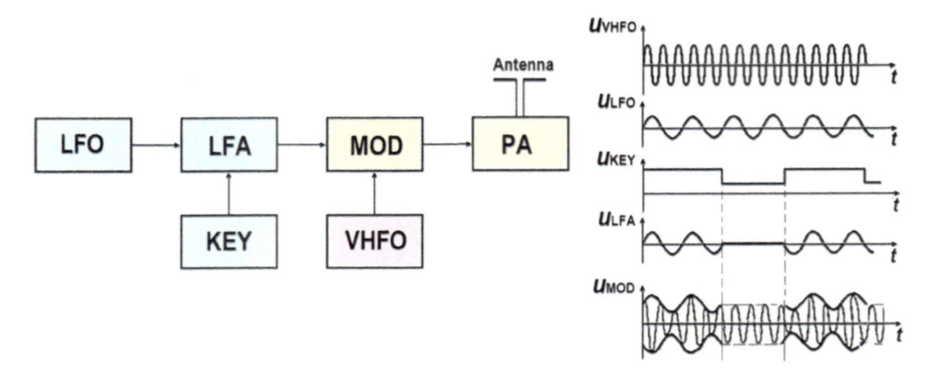

Fig. 4.31 Block diagram and waveform diagrams for a marker beacon

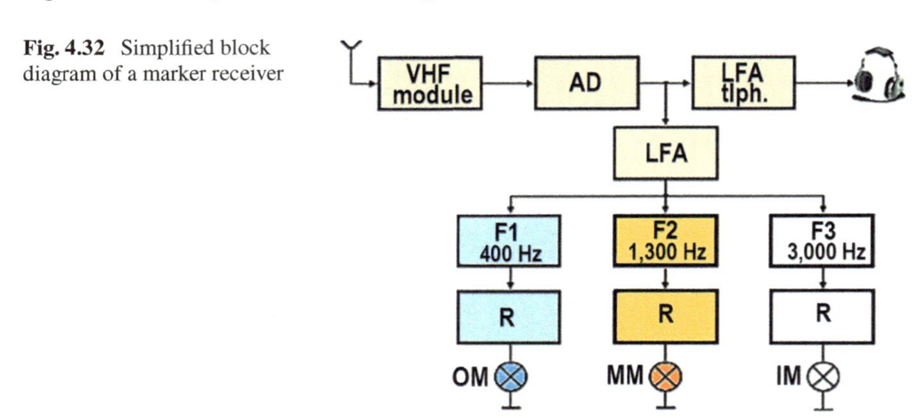

Fig. 4.32 Simplified block diagram of a marker receiver

Figure 4.32 presents a simplified block diagram of a marker receiver.

The receiver comes in superheterodyne circuit design. Signals, received by the antenna, are amplified, converted to an intermediate frequency in the VHF module of the receiver, and detected by an amplitude detector AD. After amplification in the LF amplifier (LFA), the LF oscillations at the AD output with the frequencies of 400, 1,300, or 3,000 Hz are separated by filters F1, F2, and F3 according to modulated frequencies and then enter the rectifiers R. At the rectifier outputs, the fixed voltage is formed and supplied for annunciator activation ("OM"—blue, "MM"—amber, "IM"—white).

The telephone-channel amplifier (LFA tlph) sends an output tone signal to the headphones. When the aircraft flies over the radio beacon, a tone signal of a corresponding frequency is heard in the pilot's headset (in some types of aircraft a bell rings).

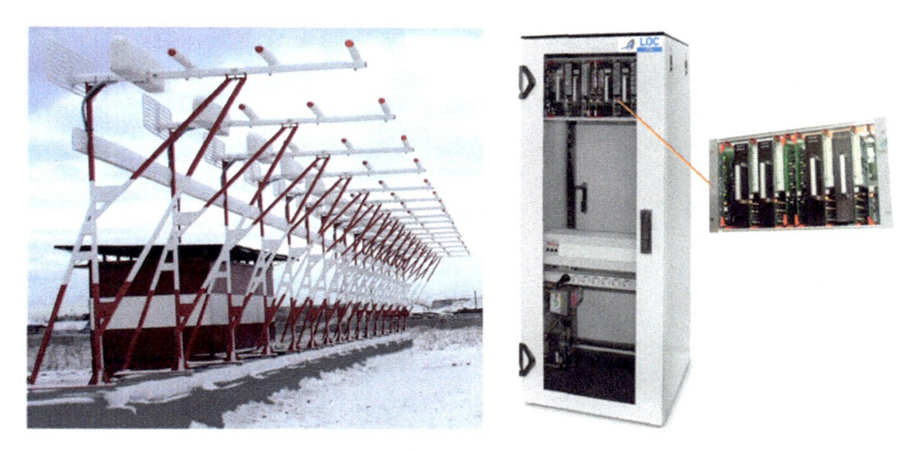

Fig. 4.33 Antenna system and the box for Loc734. *Source* http://www.nports.ru/#!/catalog/ by permission of LLC Research and development company "Radio engineering systems"

4.5 Design and Main Specifications of ILS Radio Beacons

The ILS ground equipment includes a localizer and a glide slope beacon, DME and two (three) marker beacons, localizer far-field monitoring equipment, local and remote control equipment, and equipment for monitoring the beacon radiation field parameters.

Equipment of the localizer and glide slope (glide slope and DME) beacons together with the local control equipment is in a container (operating room) which is situated near the antenna systems. The equipment for remote control of ILS beacons operation is in the ATC tower. Transmitting units of the radio beacons and the remote control equipment are mounted in boxes.

Digital methods of formation of HF and LF signals and digital HF processing of signals are widely used in modern equipment. Designers strive to decrease the overall equipment configuration due to unification of main devices and design of equipment as well as to unify its components. There is almost no wiring or its use is minimized, and power consumption is significantly reduced.

Figures 4.33, 4.34, and 4.35 present the appearance of the antennas and a ground part of equipment for ILS 734 (manufacturer—**LLC Research and development company "Radio engineering systems,"** Russia).

The following specifications should be noted for ILS:

ILS of CAT I, II and III must ensure indication of technical condition of all ground components at ATC centers during final approach of aircraft. The indication is provided by means of remote control system with required efficiency (20, 5, and 2 s for ILS of CAT I, II III, respectively).

LOC included in ILS of CAT I and II can provide operation of wireless "ground-to-air" channel (A2A emission class) simultaneously with the transmission of navigation and identification signals unless this link prevents the fulfillment of the main LOC

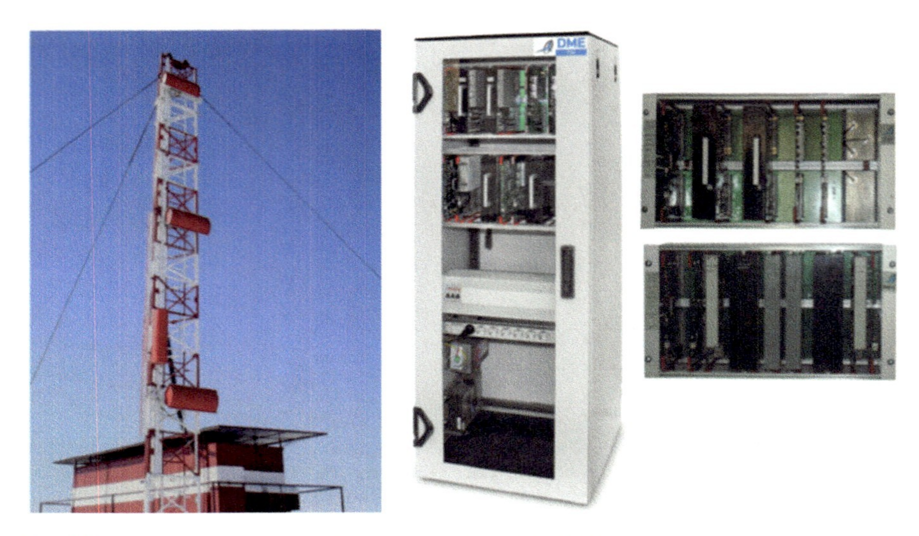

Fig. 4.34 Antenna system and the box for GP734 with DME734. *Source* http://www.nports.ru/#!/catalog by permission of LLC Research and development company “Radio engineering systems”

Fig. 4.35 The box for remote control equipment DM734, Marker734 and far-field monitoring equipment FFM734. *Source* http://www.nports.ru/#!/catalog/ by permission of LLC Research and development company “Radio engineering systems”

function. LOC included in ILS of CAT III does not provide for such a function (except in special cases).

Consider peculiarities of practical implementation of ILS using the example of ILS2700 (Russia) [4, 5].

The ILS2700 ground equipment includes Loc2700 localizer and GP + DME/NL 2700 glide path and DME radio beacon as well as remote control equipment RCE2700 normally located in the control tower. The functions performed by RCE can also be carried out by a mobile automated worksite (MAWS).

Depending on ground profile and air situation near the runway where the radio beacons are placed, ILS2700 provides parameters meeting the ICAO requirements for CAT I, II, III.

The equipment of Loc2700, GP + DME/NL2700 и RCE2700 is located in the boxes. The Loc2700 и GP + DME/NL2700 boxes include:

transmitting modules which provide formation of beacon output signals and their distribution among antenna radiating elements;

automatic devices which control the operation of radio beacons and test their parameters;

two microcomputers (for each beacon) which transmit RCE output control signals and RCE input information about the state of radio beacons. Availability of two microcomputers provides communication link redundancy;

an uninterruptible power module;

a battery module.

The transmitting and control equipment has hot standby. Switching over the standby equipment is automatic, and switchover time is no more than 2 s.

The carrier of the Loc "narrow" channel is 5.0 ± 2.2 kHz lower than the nominal frequency, and that of the "wide" channel is 5.0 ± 1.1 kHz higher that the nominal frequency.

The carrier of the GP "narrow" channel is 10.0 ± 2.2 kHz lower than the nominal frequency, and that of the "wide" channel is 10.0 ± 2.2 kHz higher that the nominal frequency.

The radio beacons operate nonstop, 24 h a day, without a constant presence of operation personnel.

A built-in test system automatically estimates the equipment state and output characteristics of the radio beacons. In the manual mode, it provides parameter measurement and diagnostics of the state of equipment down to an individual detachable device.

The radio beacons can be in the following states:

off;
standby mode;
operational mode.

Control of the radio beacons state and changeover from the standby mode to the operational one are accomplished locally or remotely.

The remote control is performed with the help of virtual buttons on the panel PC in the RCE2700 box or MAWS which can be connected to any microcomputer of the box. This is an automatic mode of the beacon operation and is considered to be the main one.

The remote control is provided only from RCE2700 box and the MAWS.

Wire links or radio channels are used for the remote control and parameter monitoring performed by RCE2700. Additional visual and audible indication of the general state of the radio beacons is provided on the information panels (included into RCE2700) located in the ATC tower.

The local control is activated by a LC/RC button on the front panel of the automatic device. The radio beacon state can be estimated due to the "Norm" and "Emergency" lights on the front panel of the automatic device.

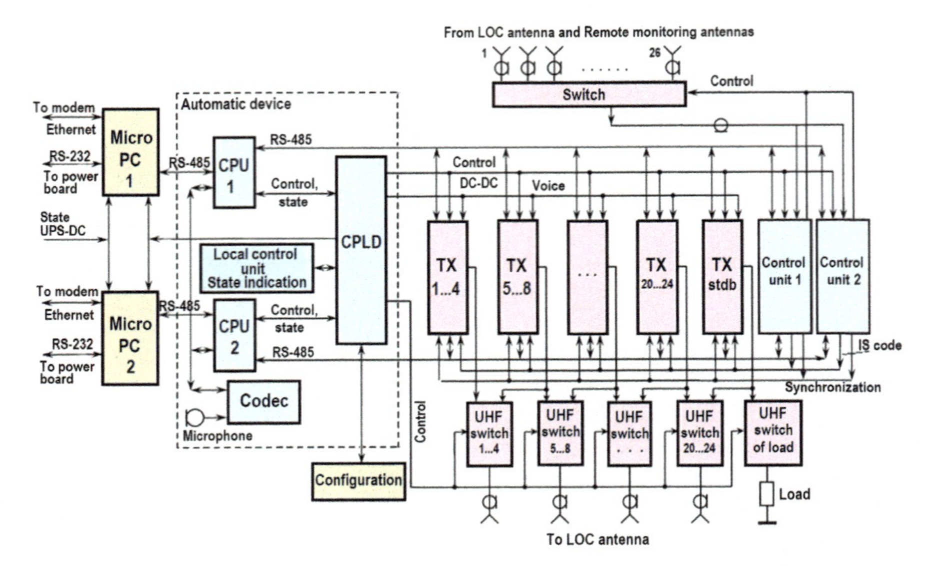

Fig. 4.36 Block diagram of Loc2700

The local control mode is not an operational one. It is designed for maintenance, e.g., troubleshooting, operations related with parameter measurements and other operations which require switching off the automatic mode. In the local control mode, the RCE2700 and the front panel of the automatic device indicate the general state of the radio beacons as emergency regardless of their actual state.

This mode provides management by means of virtual controls in the window of extended control and monitoring of the RCE2700. The virtual buttons are not active in the window of generic control.

The generalized state of a radio beacon is displayed on one or several information panels, while the RCE2700 box and MAWS provide complete monitoring and control of the radio beacon.

Loc2700 can contain 12, 16, or 24 transmitting modules (by 4 modules in a transmitter) each of which is connected to its transmitting antenna by means of UHF relay (switch) (Fig. 4.36). The GP2700 transmitter includes three transmitting modules each of which is connected to the radiating elements of a corresponding antenna by means of the UHF switch as well (Fig. 4.37). The UHF switch serves to change over the transmitting antennas from the main configuration of transmitters to the standby configuration. Each transmitter provides monitoring of the reflected and incident waves at the output of each module.

Radiation fields of Loc2700 and GP2700 are monitored by control antennas which send signals to the control units. The control units also synchronize all transmitters to provide their in-phase operation.

The Loc2700 control unit specifies an identification signal which is sent to all transmitters.

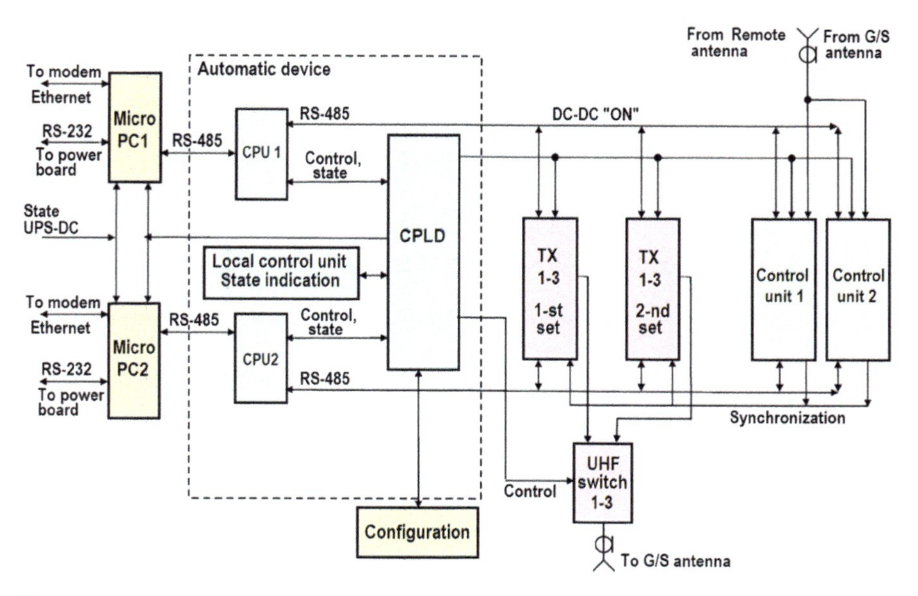

Fig. 4.37 Block diagram of GP2700

The Loc2700 automatic device includes a microphone and a digital codec to provide the voice channel. When the voice channel is activated, speech is coded by the codec and transmitted to all transmitters in a digitized form for an on-air transmission. A signal for the voice channel can be supplied to the radio beacon from outside via a communication link.

4.6 Summary

Radio-technical landing systems play important role in providing the secure and reliable aircraft's arrival on the runway. The most advanced from them are the instrumental landing systems that satisfy the most complicated categories of landing minimum. Appliance of ILS-type systems allows to increase the capacity of airfields and frequency of flights which positively influences the efficiency of aviation transport usage. Nowadays, there are no other landing systems that have the same advantages like ILS.

For the long period of ILS systems' operation their maintenance characteristics, efficiency of apparatus' control methods managed to significantly increase. In modern systems, more advanced methods of forming the field of radio beacon emission (CSB/SBO, two-channel radio beacons) are implemented that provide higher stability of glide path created by them.

However, due to the intensity and density growth of air traffic in the airfield areas, closer attention to the environment protection, and tendency to decrease the time of approaching and landing, disadvantages of ILS systems have become more and more visible. These disadvantages are limitation of radio beacons' emission sectors, creation of unique descending trajectory (glide path) which makes it impossible to implement adaptable, and in some cases more cost-effective descending and landing trajectories. Major disadvantage is the high cost of installation and operation of ILS ground-based radio beacons due to the complicity of antenna systems and their adjustment. Unfortunately, it is not possible yet to significantly decrease the influence of the external factors on the stability of glide path which requires both constant and periodical control of the emission field.

That is why, the search and development of alternative methods of landing systems that are free of mentioned disadvantages is continuing. One of the most perspective methods is the appliance of satellite navigation system for landing stage provision (this will be described in the Sect. 5.3). Quite a long and successful way has been made toward this direction. However, the possibilities of landing systems based on the satellite navigation systems [GNSS landing systems (GLS)] are lower than those of ILS on accuracy and reliability of landing provisioning.

That is why it is expected that ILS will be in operation for a long period of time and higher number of airfields will be equipped by ILS radio beacons. As long as GLS systems are improved, it is possible that ILS will be replaced by them; however, the most possible probability is the use of both systems for increasing the reliability of landing stage phase.

4.7 Further Reading

Information on construction and operating principles and main technical characteristics of Russian-made onboard and ground landing equipment can be found in [1, 2].

Information on ICAO requirements to the parameters and characteristics of ILS can be found in [3].

There is a wealth of information on ILS at https://en.wikipedia.org/wiki/Instrument_landing_system/.

References

1. Skrypnik ON (2014) Radionavigacionnye sistemy vozdushnyh sudov (Radionavigation systems of aircrafts). Moscow. INFRA-M (in Russian)
2. Hafizov AV (2014) Radiooborudovanie. CHast'2. Sredstva radionavigacii (Radio equipment: part 2. facilities of radio navigation). Kirovograd. KLA NAU. E-learning edition. Download from http://www.vipbook.su/tehnika/tehnnika/314881-radiooborudovanie-chast-2-sredstva-radionavigacii.html

3. Annex 10 to the Convention on International Civil Aviation (2006) Aeronautical Telecommunications. Volume 1. Radio Navigation Aids. 6th edn
4. Radiomayak kursovoj Loc 2700. Rukovodstvo po ehkspluatacii (Beacon Loc 2700. User manual). AECФ.461511.002PЭ. CS "Azimut" (in Russian)
5. Radiomayak glissadno-dal'nomernyj GP + DME/NL 2700. Rukovodstvo po ehkspluatacii. (beacon glide path and distance GP + DME/NL 2700. User manual). AECФ.461511.003PЭ. OAO "Azimut" (in Russian)

Chapter 5
Landing Systems Based on Satellite Navigation Systems

Global satellite navigation systems (GNSS) are the relatively new generation of radio navigation means. The differences from the traditional radio navigation means which are introduced in Chaps. 2–4 are the globality of operating area, higher accuracy, and possibility of 4D navigation tasks' solving. Modern satellite navigation systems are GPS (USA), GLONASS (Russia), Galileo (EU), and BeiDou (China).

GPS and GLONASS systems came into use in the civil aviation in the mid-end of 1990s. These systems showed their high efficiency quite quickly which led to the real revolution in the aeronautical radio navigation. New technologies of organization and air traffic management appeared and are currently evolving in the basis of GNSS as well as the technologies which provide more effective and safe usage of airspace in conditions of continuous intensity and density growth of air traffic. These technologies are reflected in the Global Air Navigation Plan (GANP) ICAO (Doc. 9750) which determines the development of global civil aviation for midterm perspective as well as in other regulatory documents of ICAO.

Unfortunately, even GNSS cannot be ideal, and they are subject to the factors that lower their efficiency. That is why new methods and functional augmentations of GNSS that allow to compensate the impact of destabilizing factors are being developed and implemented.

In accordance with the long-term forecasts, GNSS should become main and sustainable mean of aircraft's navigation on all stages of flight including approach and landing itself. That is why this chapter lays emphasis on the methods and perspectives of satellite navigation systems' usage for landing together with overview of main principles of its construction and functioning.

Section 5.1 gives common information on satellite navigation systems: Short history of development, structure and the functioning principle, and parameters of existing systems' orbit groups are analyzed. Content and differences of ephemerides information of GPS and GLONASS systems are analyzed.

GPS and GLONASS signals' structure is overviewed, and comparison analysis of these systems' signals parameters is performed. With the help of mathematical description, principles and process of solving the navigation task are described.

O. N. Skrypnik, *Radio Navigation Systems for Airports and Airways*,
Springer Aerospace Technology, https://doi.org/10.1007/978-981-13-7201-8_5

Section 5.2 analyzes main factors that affect the accuracy of positioning by satellite navigation system's signals. Main attention is focused on Geometric Dilution of Precision (GDOP) analysis. Conducted by the author experiments' results on analysis of GDOP (HDOP and VDOP), components' changes on the continuous time interval in points of observation Moscow and Irkutsk are shown.

Section 5.3 describes methods of improving the accuracy of satellite navigation system's positioning. For the case of GLONASS and GPS signals' coprocessing, results of seminatural experiments conducted by the author using Russian-made aeronautical receiver SN-4312 and imitator of signal SN-3803M are shown. Unified block diagram of differential subsystem is shown as well as the ways of realization of differential mode in functional augmentations.

At the end of Sect. 5.3, functional augmentation of GLONASS system by pseudo-lites' network is overviewed. The necessity of solving the task of pseudo-lites' positioning optimization is shown, and methods of this task's solving are described on the example of GLONASS system. Results of experiments conducted by the author on exploring the GLONASS's fields of accuracy allocation in horizontal and vertical planes in the airfields' areas at different amount of pseudo-lites are shown.

Section 5.4 describes the peculiarities of construction and main characteristics of functional augmentation of LASS local range type. Russian-made system LKKS-A-2000 is overviewed as an example of practical implementation: construction, main characteristics, block diagram, and operating principle of apparatus.

5.1 General Information of SNS

5.1.1 Performance Features and Structure of Satellite Navigation Systems

Radio navigation systems are called satellite ones if their navigation signals transmitters (radio beacons) are located on navigational satellites or space vehicles (SV) which move along fixed trajectories (orbits) around the Earth. SV can be used as radio beacons provided a user knows their coordinates and parameters of orbital motion when navigational sightings are performed.

Values of parameters of SV orbital motion calculated for fixed points in time according to results of its motion forecasting are known as ***ephemerides***. A set of ephemerides for all satellites of a system is known as an ***almanac***.

The development of satellite navigation systems (SNS) is based on experiments carried out in 1957–1958. They were devoted to determination of motion variables of a man-made earth satellite on the results of Doppler measurements of its radiated signals. At the same time, it became possible to solve the opposite problem: To fix a receiver using the Doppler shift of a signal radiated by a satellite provided, the coordinates of the satellite are known.

This principle of navigational sightings formed the basis for the first generation of SNS. Those SNS were TRANSIT (NAVSAT, Navy Navigation Satellite System) (1964, the USA), the world's first navigation satellite system, and its analog CIKLON (CIKADA-M) (1967, the SU).

First of all, the systems were designed for submarine navigation and navigational support of submarine-launched ballistic missiles, but later they were used for commercial purposes as well.

The peculiarity of the first-generation SNS was the use of satellites placed in low polar orbits. The height of the circular orbits above the Earth surface was approximately 1,100 km, and satellite period was approximately 107 min. With those orbit parameters, an SV was in sight of a user no more than for 40 min.

The orbit group included 6–7 satellites, and only one satellite was usually visible at any given time. To fix the user position, Doppler measurements were performed for signals received from a satellite. The law of variation of the Doppler shift depends on the position and motion variables of the satellite which should be known and on receiver coordinates. On the basis of those, the unknown user position was fixed through rather complicated calculations and successive iterations, while several Doppler measurements were needed.

Such a procedure provided a user position fix with 1–2 h discretization at mid-latitudes and several hours at the equator. This fact as well as low (worse than 200 m for highly dynamic users) accuracy of coordinate measurements was the main significant disadvantages of the first-generation SNS.

To provide continuity and high accuracy of navigational sightings, it was necessary to increase the time of satellite visibility and the number of satellites for simultaneous measurements as well as to use other principles of position fixing. This could be done by increasing the orbit radius and the number of orbit group constituents, by using other types of signals and methods for determination of navigational parameters.

In addition to the above, a number of problems had to be solved:

to provide mutual synchronization of signals (time scales) of the satellites with the necessary accuracy;
to increase the accuracy of determination and forecasting of SV orbit parameters;
to ensure separate reception of signals from the satellites monitored simultaneously;
to ensure reception and processing of low-power signals from satellites against the background of natural noises near the ground surface.

The mentioned problems were solved in the second-generation SNS, that is Global Positioning System (GPS) (1994, the USA) and *GLObalnaya NAvigacionnaya Sputnikovaya Sistema (GLONASS)* (1995, the SU/Russia).

In the early stages of developing and putting into service, another name of GPS, *NAVigation Satellites providing Time And Range* (NAVSTAR), was more common.

The developed technical solutions and achieved physical and operational characteristics of the second-generation SNS have been so successful that the systems are continuing to develop nowadays embracing new areas of application. Besides GPS and GLONASS, such global systems as GALILEO (European Union) and BeiDou

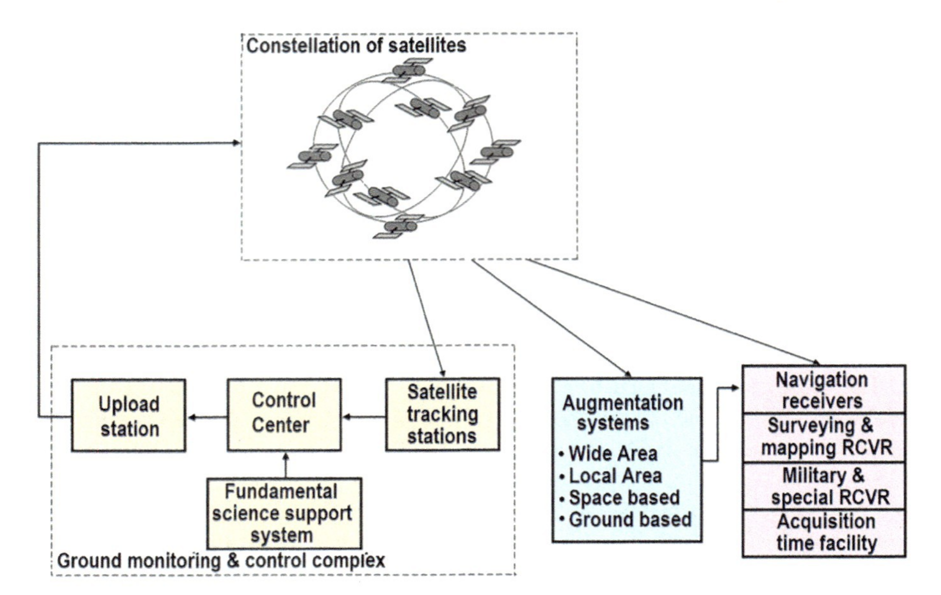

Fig. 5.1 Constituents of a satellite navigation system

(Compass) (China) as well as regional systems Indian Regional Navigation Satellite System (IRNSS) and Quasi-Zenith Satellite System (QZSS, Japan) are currently being introduced into service. In the medium term (until 2030), SNS, which are generally referred to as Global Navigation Satellite Systems (GNSS), can become a primary means of aircraft navigation.

Modern GNSS are designed to determine three spatial coordinates (latitude, longitude, and altitude), three constituents of a space velocity vector and the precise time of different users. SNS is a complex of interacting radio-electronic means installed on the satellites and ground stations as well as user equipment (receivers) of various purposes (Fig. 5.1).

The basis of the system is a network (constellation) of satellites serving as movable radio beacons—sources of navigational signals. The satellites are in fixed near-circular Medium Earth Orbits (MEO) at the altitudes of between 19,000 and 24,000 km from the Earth's surface.

A ground part of the system is composed of a ground complex for system control and its continuity monitoring. It serves to determine deviations of actual parameters of satellites' orbital motion, parameters of Earth rotation, and other parameters (constants) from the nominal values used in system control and navigational sightings, to form and input corresponding corrections to the system.

The control and monitoring complex includes a center of system control, information collection and processing, stations for satellite monitoring and trajectory measurements upload stations, and stations for navigation-and-time field monitoring. The ground facilities also include a complex of augmentations increasing the system accuracy for definite groups of users, a complex of fundamental (scientific) support of system operation.

The equipment for users of navigational and time information includes receivers of varying purposes (first of all, for civil and military use) which are notable for their accuracy, interface for interaction with the user or other systems, an option set and service functions (airborne, marine, geodesic, home, and other receivers).

Consider the main peculiarities of SNS which distinguish them from other technical means of navigation:

1. Location of transmitting stations on the satellites which move with considerable speeds, thus leading to a significant Doppler shift. The frequency shift is also influenced by gravitational effects.
2. Use of the SHF band for measurements of navigational parameters (for passing a signal through ionospheric layers);
3. Coverage of the entire Earth surface (globality of the working area) and a very high (single meters–a few tens of meters) positioning accuracy for highly dynamic users;
4. Low-energy performance of satellite-to-user radio links due to considerable distances travelled by radio signals and limitations of the mass and dimensions of the transmitting equipment which decreases the power of a radiated signal;
5. Necessity of ephemeris information for user position fixing and necessity of periodic correction of the ephemeris information.

5.1.2 Parameters of Orbit Groups and the Contents of Ephemeris Information

The orbital movement of man-made Earth satellites is described using Kepler's laws. Parameters defined the orbit shape, and its position in space and the position of a satellite in it are known as Keplerian orbital elements.

The orbit of a man-made Earth satellite is plane and generally elliptic. Point P in the orbit which is nearest to the foci is known as a ***perigee,*** while point A in the orbit which is farthest from the foci is known as an ***apogee***. A line connecting an apogee with a perigee is known as an ***apse line***.

The orientation of orbit relative to the Earth is characterized by (Fig. 5.2):

inclination i of orbit, i.e., an angle between the equatorial plane and the orbital plane;
longitude of the ascending node $\boldsymbol{\Omega}$, i.e., the longitude L_{Ω} of a point where the orbital plane intersects the equator and where the satellite moves from the southern hemisphere to the northern one. This point is known as an ascending node;
argument of periapsis ω, i.e., an angle between a node line (line of intersecting the orbital and equatorial planes) and an apse line.

The orbit form (i.e., degree of its ellipticity) is characterized by the ***eccentricity***

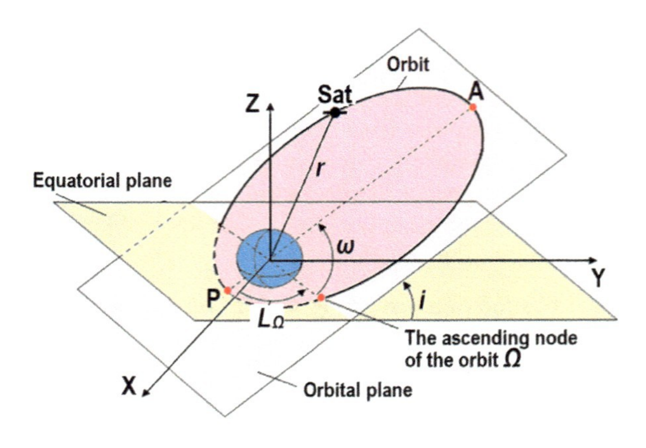

Fig. 5.2 Orbital parameters of the Earth satellite

$$e = \frac{\sqrt{a^2 - b^2}}{a},$$

where a is a semimajor axis, and b is a semiminor axis of an elliptic orbit.

Position of a satellite in the orbit is characterized by the length of ***radius vector r*** and ***true anomaly υ***, i.e., an angle between the apse line and the radius vector which is related according to expression

$$r = \frac{a(1 - e^2)}{1 + e \cos \upsilon}.$$

The speed of a satellite relative to the Earth is a function of height and true anomaly. It is constant for a circular orbit and decreases as the altitude of the satellite over the Earth rises.

If a satellite moves in a circular orbit at the height of 35,810 km in the direction coinciding with the Earth rotation, it appears to hover over a fixed point of the equator (its speed relative to the earth surface is equal to zero) provided $i = 0$. A satellite like this is called ***GEO*** (geostationary).

Satellites rotating in orbit with the period equal to or multiple of an equinoctial day (23 h 55 min 56.6 s) are called synchronous ones (***GSO***).

Parameters describing orbit groups of satellite navigation systems are shown in Table 5.1.

As at December 25, 2018:

GLONASS constellation contains 26 satellites, 24 of which are used for the intended purpose, one is taken out of service for spares, and one is at the stage of flight testing.
GPS constellation contains 32 satellites, 31 of which are used for the intended purpose and one is taken out of service for maintenance.
BeiDou constellation contains 38 satellites, 17 of which are used for the intended purpose (included in operational orbital constellation).

Table 5.1 Parameters of SNS orbit groups

Parameter	GPS	GLONASS	Galileo	BeiDou
Number of satellites (nominal), pcs	24	24	30	35 (27—MEO, 3—GSO, 5—GEO)
Number of orbits, pcs	6—MEO	3—MEO	3—MEO	3—MEO, 3—GSO
Orbital height, km	20,145	19,100	23,222	21,500 35,786
Orbit type	Circular	Circular	Circular	Circular, GSO, GEO
Inclination of orbit, deg.	55	64,8	56	55
Satellite period	11 h 56 min 54 s	11 h 15 min 44 s	14 h 4 min 42 s	12 h 38 min
System of coordinates	WGS-84	PZ-90.02	GTRF	CGCS2000
Orbit shift in longitude of ascending node, deg.	60	120	120	120—for MEO 0—for GSO

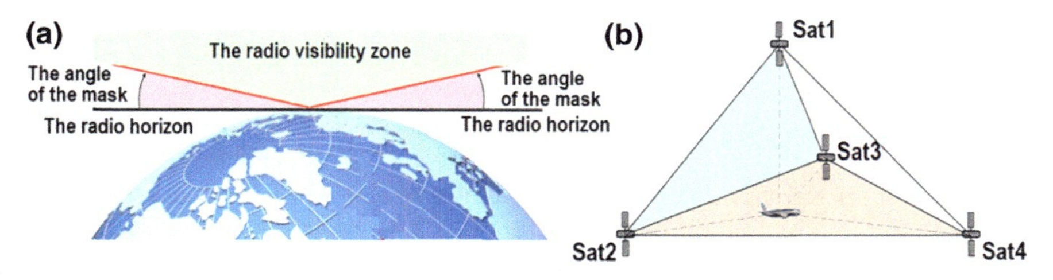

Fig. 5.3 Concerning the radio-visibility zone and optimal working constellation

GLONASS and GPS use different ephemeris information for describing the satellite position in orbit and forecasting its motion (Table 5.2).

The navigational message transmitted by satellites also includes other information (data about the satellite state, parameters of the Earth model, ionosphere, almanac age, and so on).

The user equipment receives signals only from satellites which are within the radio-visibility zone, i.e., above the radio horizon (Fig. 5.3a). To eliminate faults of the receiving equipment due to instability of signal propagation close to the radio horizon, only the satellites above the mask angle (nominal value $\alpha = 5^{\circ}$) are processed (are included in the working constellation).

From 7 to 12–14 satellites can be in the radio-visibility zone at a time.

There is a concept of optimal working constellation in terms of the highest accuracy of positioning. For an optimal working constellation of four satellites, the volume

Table 5.2 Contents of GPS and GLONASS ephemeris information

GPS	GLONASS
PRN—pseudo-range number	N_S—total number of satellites transmitting CDMA signals (1–63) which are referenced to the almanac
Date—base date (UTC)	Date—day number (1–1,461) within a four-year interval starting on January 1 of the last leap year according to Moscow decree time (UTC + 3 h) [DD MM YY]
t—reference time, s	T_{Ω}—time of the day when the satellite is crossing its first orbital node (from 00 h 00 m 00 s base date), s
e—eccentricity	e—eccentricity
i—orbital inclination, deg.	i—adjustments to nominal inclination (64.8°) of the satellite orbit at the moment of ascension
$d\Omega/dt$—rate of right ascension W, deg/s	T_{rot}—adjustments to the satellite's nominal draconic orbital period at the moment of ascension, s
A—semimajor axis, km	n_l—number of frequency slug
L_{Ω}—longitude of ascending node (deg.) on 00 h 00 min 00 s base date	L_{Ω}—longitude of the satellite's first orbital node, deg.
ω—argument of perigee, deg.	ω—argument to satellite's perigee at the ascension time, deg.
$af0$—clock correction, s	$\delta t2$—rough correction from onboard time scale to the GLONASS time scale, s
$af1$—rate of clock correction $af0$, s/s	ΔT—speed of change of the draconic orbital period at the moment of ascension, s/s
m—mean anomaly, deg.	–

of the tetrahedron made up by the satellites is maximal (Fig. 5.3b). One satellite must be in the zenith, and the others make up an equilateral triangle in the horizon plane. The position-fixing object is in the same plane as well.

5.1.3 Signal Structure

GPS and GLONASS are dual-use systems—they are designed for military and civil users. So, they provide for special measures allowing military users to fix position more accurately.

Implementation of the measures is ensured by structure and parameters of navigation signals transmitted by satellites as well as by an opportunity for different users to access these signals.

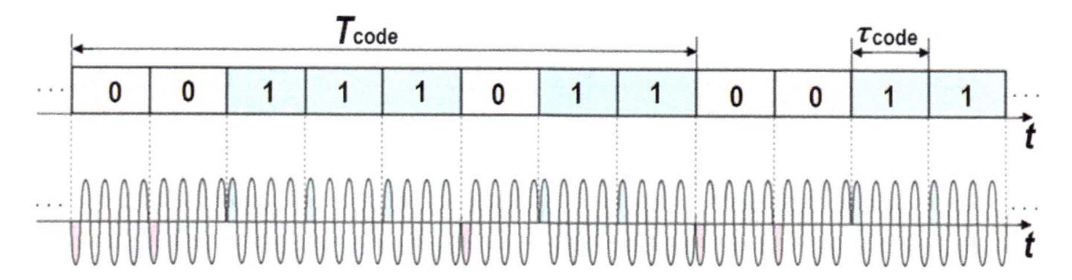

Fig. 5.4 Formation of a navigation signal

SNS uses a no-request range measurements between navigational satellite and the user. To measure the distance to the source of signal radiation–satellite, it is necessary to know the time of signal radiation and to measure the time of its reception. The time scales of the transmitter (satellite) and receiver (user) should be synchronized.

If single radio pulses or their packs (sequences) are radiated, it is impossible to detect and extract this signal against the background of natural noises because of significant attenuation when propagating from the satellite to the earth surface. Besides, it is impossible to transmit additional ephemeris information.

Special wideband navigation pseudo-random noise (PRN) signals are therefore applied. PRN is a signal with the spectrum similar to the spectrum of white Gaussian noise. The signals are formed due to binary phase-shift keying (BPSK) with a special distance-measuring binary code of a harmonic signal (carrier) (Fig. 5.4). The distance-measuring code is a pseudo-random periodic sequence of pulses, e.g., maximum length sequence, Gold code. The carrier signal is generated by the satellite time-and-frequency standard (atomic frequency standard).

The distance-measuring code called a navigation code has code element duration τ_{code} and is formed continuously with period of element code repetition T_{code}. Correspondingly, a navigation signal is radiated by the satellite continuously as well.

Each symbol of the distance-measuring code is "tied" to the time scale produced by the satellite time-and-frequency standard (a highly stable generator). This time scale is, in its turn, synchronized with the UTC.

So, on the receiver side, we can know, for each specific code element, the time of its radiation by the satellite equipment using the UTC time scale. For this purpose, the time scale created in the receiver should be synchronized with the time scale of the satellite equipment.

Thus, the distance between a satellite and a user is determined by measuring the time of propagation (delay) of a distance-measuring code.

A spread-spectrum signal of a satellite can be detected and received even if its level is lower than that of natural noises in the reception point. It is provided due to a special algorithm of receiving and processing a signal from a satellite received along with the noise. A device which implements the algorithm is known as a ***correlator***.

Element-by-element, the correlator compares implementations of a received signal and noise with a signal copy formed in the receiver itself. This results in formation of a function of their mutual correlation. The signal pattern is shifted along the time

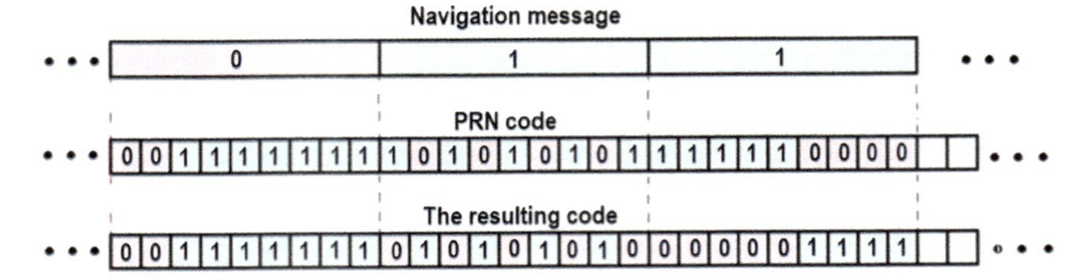

Fig. 5.5 Resultant navigation code

scale of the receiver until the correlation function reaches its peak. Meanwhile, the total energy of received signal elements gets larger than the noise level, and the signal can be detected and processed to extract the desired information.

To fix the user position according to measured delays of received signals, it is necessary to know ephemerides of satellites at the time of solving the navigational problem. A navigation signal radiated by each satellite contains therefore not only navigation but also operation information. The operation information in form of a navigational message includes data about ephemerides of the satellite as well as data about the almanac of the system.

The operation information is also transmitted with binary symbols 0 and 1, but their duration is much greater than that of distance-measuring code symbols.

In the transmitter of a satellite, the operation information symbols are added modulo 2 with the distance-measuring code symbols, thus resulting in creation of a digital code with which the carrier signal is phase-shift keyed (Fig. 5.5).

The operation information is transmitted as superframes with the speed of 50 bit/s. A superframe contains information about ephemerides of the satellite and a complete almanac of the system. Duration of a superframe is 2.5 min in GLONASS and 12.5 min in GPS. A superframe consists of 5 frames with the duration of 30 s. Each frame contains 15 lines of information with the duration of 2 s. The lines provide information about ephemerides of the satellite, a part of the almanac and check symbols.

At its input, the user receiver gets signals from all satellites which are within the visibility zone. To be distinguished in the receiver, the signals must differ in some parameters.

In GPS, a code-division multiple access (CDMA) is used for distinguishing signals from different satellites. All satellites radiate signals at the same frequency, but a distance-measuring code is individual for each satellite.

GLONASS uses a frequency-division multiple access (FDMA). All satellites radiate signals with the same distance-measuring code, but each radiates at a specific carrier frequency.

Accuracy of position fixing depends upon accuracy of measuring the distance to the satellite. The accuracy of measuring the distance is greatly affected by:

an additional unknown delay of a signal when propagating through the ionosphere and troposphere;
a kind of the used signal;
duration of a distance-measuring code element.

In the ionosphere, the trajectory of a signal distorts, which leads to an additional unknown time delay of its propagating. The signal delay in the ionosphere, however, depends on the signal frequency. So, if the signal is received at two different frequencies, the additional unknown delay can be calculated and compensated in the receiver.

Military receivers can operate at two frequencies, others cannot. To make the signal at the second frequency inaccessible for conventional receivers, it contains a special distance-measuring code which cannot be applied without knowing the key.

In GLONASS, a ***standard-precision signal*** (signal of SP code) accessible for all users is transmitted at the frequencies of 1,598–1,605 MHz (L1 band). In GPS, a ***clear access signal*** (signal of C/A (coarse/acquisition) code) is transmitted at the frequency of 1,575.42 MHz (L1 band).

In GLONASS, an obfuscated ***high-precision signal*** (signal of HP code) accessible only for military users is transmitted at two frequencies 1,598–1,605 MHz (L1 band) and 1,242–1,248 MHz (L2 band).

In GPS a ***precise signal*** (signal of P (protected) code) is transmitted at the frequencies of 1,575.42 MHz (L1 band) and 1,227.60 MHz (L2 band).

Meanwhile, duration of HP (P code) element is ten times less than duration of SP (C/A code) element.

Table 5.3 Parameters of GLONASS and GPS signals

Parameter	GLONASS	GPS
Channel separation	FDMA	CDMA
Duration of a code element, μs		
SP (C/A)	2	1
HP (P)	0.2	0.1
Period of code, elements		
SP (C/A)	511	1,023
HP (P)	511,000	7 days
Radiation frequency, MHz		
SP (C/A) and HP (P)	1,598–1,605	1,575.42
HP (P)	1,242–1,248	1,227.60
Speed of transmission of operation information, bit/s	50	
Frame duration, s	30	
Superframe duration, min	2.5 (5 frames)	12.5 (25 frames)
PRN type	MLS	Gold code
Polarization	Right-handed circular	

The main parameters of GLONASS and GPS signals are shown in Table 5.3.

Notice that elements of distance-measuring codes and operation information code are formed, like a carrier frequency signal, by the same source—a highly stable satellite time-and-frequency standard (atomic frequency standard).

5.1.4 Principles of Solving a Navigational Problem

To determine three space coordinates (x, y, z) and three constituents of a velocity vector (V_x, V_y, V_z), a SNS uses a distance-measuring and Doppler method.

Implementation of the distance-measuring method requires production (in the user receiver), a time scale synchronized with the SNS system time.

The measured value of the distance to ith satellite in the geocentric coordinate system is determined by expression

$$D_i = \sqrt{(x_i - x)^2 + (y_i - y)^2 + (z_i - z)^2} + c\Delta t_{\mathrm{ai}} + c\Delta\tau_{\mathrm{syn}}, \quad (5.1)$$

where x_i, y_i, and z_i are known coordinates of ith satellite; x, y, and z are desired user coordinates; $c\Delta t_{\mathrm{ai}}$ is an additional signal delay due to conditions of radio-wave propagation, first of all, due to trajectory curvature during ionospheric and tropospheric refraction; and $c\Delta\tau_{\mathrm{syn}}$ is an unknown value caused by an error of time-scale synchronization.

The influence of $c\Delta t_{\mathrm{ai}}$ summand is compensated automatically in receivers capable of receiving signals at two frequencies, so we will rule it out of the further analysis.

The range measured with the error of time-scale synchronization $\Delta\tau_{\mathrm{syn}}$ is known as ***pseudo-range***.

Equation (5.1) includes four variables (user coordinates and synchronization error) for pseudo-range. For their unambiguous determination, it is necessary to have four linearly independent equations of (5.1) type, i.e., to measure pseudo-ranges to four satellites simultaneously. So, the minimal for solving the navigational problem number of satellites within the visibility zone is equal to four.

The number and placement of satellites in GPS and GLONASS orbit groups are chosen so that at least four satellites can be seen at any point of earth surface at any time.

An equation for radial speed (rate of range change) between the ith satellite and the user is written as

$$V_{\mathrm{ri}} = \frac{1}{D_i}\left[(x_i - x)(V_{xi} - V_x)(y_i - y) + \left(V_{yi} - V_y\right) + (z_i - z)(V_{zi} - V_z)\right] + V_{\mathrm{r\,syn}}. \quad (5.2)$$

Equation (5.2) includes known (from the ephemeris information) coordinates and constituents of satellite speed and unknown coordinates (x, y, z), constituents of user speed (V_x, V_y, V_z), and a summand $V_{r.syn}$ caused by the rate of synchronization error change.

The user coordinates can be found from expression (5.1). So, only the constituents of user speed and the rate of synchronization error change stay unknown. To find them, it is necessary to have four linearly independent equations of (5.2) type, i.e., to measure radial speeds to four satellites.

The radial speed of the user relative to the ith satellite is determined by measuring ΔF_{Di} Doppler shift of the signal received from it

$$V_{\mathrm{ri}} = \lambda \Delta F_{\mathrm{Di}}/2,$$

where λ is the wavelength of oscillations.

In practice, the modern user equipment solves a navigational problem with the account of all visible satellites. It ensures some increase in accuracy of navigational sightings. At first, the coordinates and constituents of user speed in Cartesian coordinate system [expressions (5.1) and (5.2)] are determined, and then, they are converted into geodesic coordinates and the altitude (expressions in Sect. 1.4).

5.2 Factors Influencing SNS Accuracy

The accuracy of coordinate and speed constituent determination according to SNS data depends on many factors. The main of them are:

discrepancy of ephemerides and actual parameters of satellite motion. This leads to use of inaccurate values of variables in Eqs. (5.1) and (5.2), and hence, to inaccuracy of solving the systems of equations relative to unknowns.
errors of time-and-frequency provision. This leads to a shift of the satellite time scale relative to the UTC and an additional error in summands $c\Delta\tau_{\mathrm{syn}}$ and $V_{\mathrm{r\,syn}}$ in expressions (5.1) and (5.2).
conditions of radio signal propagation in the ionosphere and troposphere. This leads to an additional delay of a signal when propagating along the radio path due to trajectory curvature [the second summand in expression (5.1)].
multipath propagation of radio signals. The receiving antenna receives signals re-reflected from ground features, so the receiver works out parameters of the sum signal.
relative position of a user and satellites according to which navigational sightings are performed. It influences the angle of intersecting the position lines (lines of equal distances) affecting the accuracy of fixing the position using the distance-measuring method.

Besides the factors mentioned above, the SNS accuracy is influenced by imperfection of equipment located at satellites and user equipment (instability of power

Table 5.4 Values of main constituents of the pseudo-range measurement error [1]

Error, σ, m	Zenith satellite	Horizon satellite
Ephemeric	1.0	2.4
Time-and-frequency provision	2.1	2.1
Tropospheric	0.3	1.5–3.0
Ionospheric	3–15	10–50
Multipath	–	0–3.0
Imperfection of equipment, external noises	2.0	3.0–6.0
Total (according to monitoring systems)	4.5–9.2 (GLONASS) 3.3–8.5 (GPS)	

sources, thermal noises of radio elements, intra-system disturbances, algorithms of signal processing and circuit designs implemented in receivers, and so on).

Table 5.4 shows approximate values of errors which make the mentioned above factors a part of the resultant error of pseudo-range measurement in GLONASS.

As can be seen from Table 5.4, some errors are smaller for the zenith satellite than for the horizon one. This is because the horizon satellite has a longer radio path in ionospheric and tropospheric layers as well as above the earth surface.

To estimate the effect of relative position of a user and satellites on the SNS accuracy, a concept of ***geometric dilution of precision*** (***GDOP***) or DOP is used.

An expression for GDOP can be obtained by expansion of (5.1) in Taylor series in the neighborhood of true values of variables. Meanwhile, we will not consider $c\Delta t_{\mathrm{ai}}$ summand as it can be compensated in two-frequency receivers. Limited to linear terms of the series, we obtain

$$D_i = D_{i0} + \left.\frac{\partial D_i}{\partial x}\right|_{\substack{x=x_0\\y=y_0\\z=z_0}} \delta x + \left.\frac{\partial D_i}{\partial y}\right|_{\substack{x=x_0\\y=y_0\\z=z_0}} \delta y + \left.\frac{\partial D_i}{\partial z}\right|_{\substack{x=x_0\\y=y_0\\z=z_0}} \delta z + c\Delta\tau_{\mathrm{syn}}, \tag{5.3}$$

where $\delta D_i = D_i - D_{i0}$ is an error of ith satellite range measurement; D_{i0} is a true range to the ith satellite; x_0, y_0, and z_0 are true user coordinates; and δx, δy, and δz are errors of fixing the user position.

To fix the user position, it is necessary to measure ranges to four satellites, so we will introduce vector $\mathbf{\delta Z}^{\mathrm{T}} = |\delta D_1, \delta D_2, \delta D_3, \delta D_4|$ which includes errors of measuring the ranges to these satellites.

We will also introduce vector $\mathbf{\delta X}^{\mathrm{T}} = |\delta x,\ \delta y\ , \delta z\ , \Delta\tau_{\mathrm{syn}}|$ which includes errors of user coordinate measurements and the error of synchronization of the receiver time scale with the UTC.

Then, expression (5.3) can be written as a vector-matrix one

$$\boldsymbol{\delta}\mathbf{Z} = \boldsymbol{H}\boldsymbol{\delta}\mathbf{X}, \tag{5.4}$$

where $\boldsymbol{H}$ is a matrix of direction cosines of user–satellites range lines in *OXYZ* coordinate system which is written as

$$\boldsymbol{H} = \left| \begin{array}{cccc} \frac{x_1-x}{D_1} & \frac{y_1-y}{D_1} & \frac{z_1-z}{D_1} & 1 \\ \frac{x_2-x}{D_2} & \frac{y_2-y}{D_2} & \frac{z_2-z}{D_2} & 1 \\ \frac{x_3-x}{D_3} & \frac{y_3-y}{D_3} & \frac{z_3-z}{D_3} & 1 \\ \frac{x_4-x}{D_4} & \frac{y_4-y}{D_4} & \frac{z_4-z}{D_4} & 1 \end{array} \right|. \tag{5.5}$$

From expression (5.4), a matrix of variances of range measurement errors $\boldsymbol{D}_{\delta Z} = D\{\boldsymbol{\delta}\mathbf{Z}\} = M\{\boldsymbol{\delta}\mathbf{Z}\boldsymbol{\delta}\mathbf{Z}^{\mathrm{T}}\}$ can be found where "T" is a transposition symbol, $M\{\cdot\}$ is a mathematical expectation symbol, and $D\{\cdot\}$ is a dispersion symbol. Taking into account conjugation rules, we obtain

$$\boldsymbol{D}_{\delta Z} = M\{\boldsymbol{H}\boldsymbol{\delta}\mathbf{X}\boldsymbol{\delta}\mathbf{X}^{\mathrm{T}}\boldsymbol{H}^{\mathrm{T}}\} = \boldsymbol{H}M\{\boldsymbol{\delta}\mathbf{X}\boldsymbol{\delta}\mathbf{X}^{\mathrm{T}}\}\boldsymbol{H}^{\mathrm{T}} = \boldsymbol{H}\boldsymbol{D}_{\delta X}\boldsymbol{H}^{\mathrm{T}}.$$

From this, we obtain

$$\boldsymbol{D}_{\delta \mathrm{X}} = \boldsymbol{H}^{-1}\boldsymbol{D}_{\delta \mathrm{Z}}(\boldsymbol{H}^{\mathrm{T}})^{-1},$$

where (-1) is a matrix inverse symbol.

Considering the range measurements to all satellites as uniformly precise $(\sigma_{\delta D1} = \sigma_{\delta D2} = \sigma_{\delta D3} = \sigma_{\delta D4} = \sigma_{\delta D})$, we obtain

$$\boldsymbol{D}_{\delta X} = \boldsymbol{H}^{-1}\left(\boldsymbol{H}^{\mathrm{T}}\right)^{-1}\sigma_D^2 = \left(\boldsymbol{H}^{\mathrm{T}}\boldsymbol{H}\right)^{-1}\sigma_D^2. \tag{5.6}$$

The accuracy of space–time determinations of a user can be defined as a sum of diagonal elements of matrix $\boldsymbol{D}_{\delta X}$. Then

$$\sigma_{\delta x}^2 + \sigma_{\delta y}^2 + \sigma_{\delta z}^2 + \sigma_{\Delta\tau_{\mathrm{syn}}}^2 = \mathrm{tr}\left[\left(\boldsymbol{H}^{\mathrm{T}}\boldsymbol{H}\right)^{-1}\right]\sigma_D^2 = \mathrm{GDOP}\cdot\sigma_D^2,$$

where

$$\mathrm{GDOP} = \sqrt{\mathrm{tr}\left[\left(\boldsymbol{H}^{\mathrm{T}}\boldsymbol{H}\right)^{-1}\right]} \tag{5.7}$$

is a coefficient which correlates the error variance of space–time determinations and the error variance of measuring the range to satellites.

The following constituents of GDOP are distinguished:

horizontal **(HDOP)** defines the accuracy of position fix in horizontal plane;
vertical **(VDOP)** defines the accuracy of position fix in vertical plane;

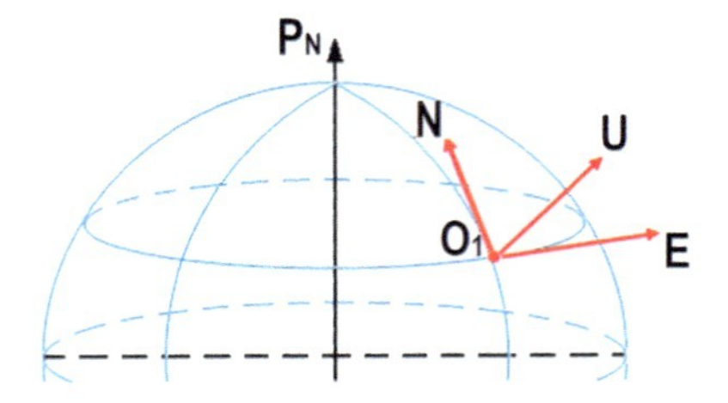

Fig. 5.6 Local coordinate system

position (**PDOP**)	defines the accuracy of position fix in space;
timing (**TDOP**)	defines the accuracy of a correction to the user time scale.

The overall factor

$$\mathrm{GDOP}^2 = \mathrm{PDOP}^2 + \mathrm{TDOP}^2, \tag{5.8}$$

where $\mathrm{PDOP}^2 = \mathrm{HDOP}^2 + \mathrm{VDOP}^2$ and it is common for SNS that VDOP > HDOP, i.e., the accuracy of altitude determination is poorer than that of horizontal coordinates.

GDOP and its constituents calculated with the use of expressions (5.5)–(5.8) define the accuracy of position fix in the coordinate system $O_1X_1Y_1Z_1$ with origin O_1 coinciding with the user position, while axes are parallel to those of the rectangular geocentric coordinate system *OXYZ*. For practice, it is necessary to determine VDOP and HDOP in the local coordinate system O_1UNE with U axis pointed to the geodesic zenith of point O_1, N axis to the north and E axis to the east (Fig. 5.6).

Conversion of point coordinates from *OXYZ* system to O_1UNE local coordinate system meets the expression

$$\begin{vmatrix} x_{\mathrm{UNE}} \\ y_{\mathrm{UNE}} \\ z_{\mathrm{UNE}} \end{vmatrix} = \boldsymbol{M} \begin{vmatrix} x - x_0 \\ y - y_0 \\ z - z_0 \end{vmatrix},$$

where x, y, and z are satellite coordinates, and x_0, y_0, and z_0 are coordinates of point O_1 in the rectangular geocentric coordinate system. Using the coordinate conversion matrix

$$\boldsymbol{M} = \begin{vmatrix} -\sin L_0 & \cos L_0 & 0 \\ -\cos L_0 \sin B_0 & -\sin L_0 \sin B_0 & \cos B_0 \\ \cos L_0 \cos B_0 & \sin L_0 \cos B_0 & \sin B_0 \end{vmatrix}$$

where L_0, B_0 are geodesic coordinates of point O_1, the expression for GDOP in O_1UNE coordinate system can be written as

$$\mathrm{GDOP}_{\mathrm{UNE}} = \sqrt{\boldsymbol{M}(\mathrm{GDOP}^2)\boldsymbol{M}^{\mathrm{T}}}. \tag{5.9}$$

Table 5.5 Dependence of DOP on the number of visible satellites [1]

Parameters	Number of visible satellites (*N*)				
	4	5	6	7	8
P_N	1	1	1	1	0.91
HDOP	1.41	1.26	1.15	1.03	0.95
VDOP	2.0	1.75	1.7	1.61	1.6
TDOP	1.13	1.03	1.03	0.95	0.93
PDOP	2.45	2.16	2.05	1.91	1.86
GDOP	2.69	2.39	2.3	2.13	2.08

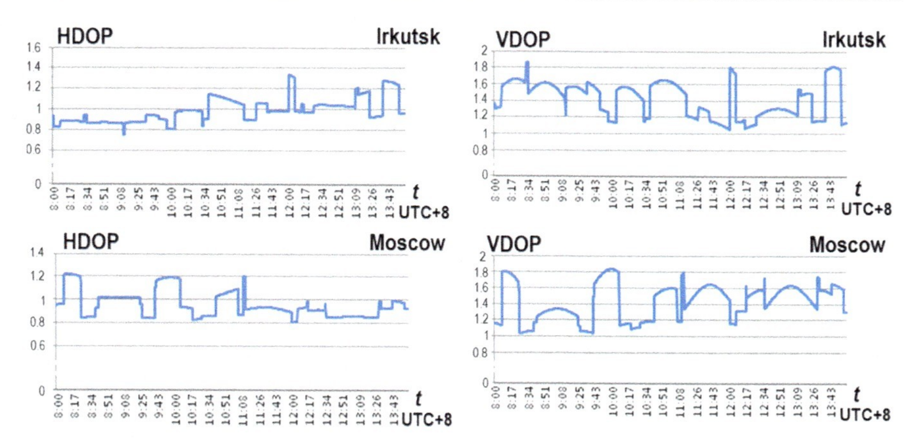

Fig. 5.7 Change of HDOP and VDOP in time

DOP value depends on the number of satellites within the visibility zone and their location relative to the position-fixing object. Table 5.5 contains values of GDOP constituents for a different number of visible satellites.

As follows from Table 5.5, an increase in number of visible satellites leads to decrease in GDOP constituents, i.e., the accuracy of SNS time–space determinations rises. But as the number of visible satellites increases, the relative benefit for increase in accuracy of space–time determinations falls. For the existing constitution of GLONASS and GPS orbit groups, probability of presence of at least seven satellites within the visibility zone is a true event provided there is no shadow effect.

Motion of satellites in orbits causes a continuous GDOP change in the point of the user position. As an example, Fig. 5.7 shows HDOP and VDOP changes for GLONASS system in two points (Irkutsk and Moscow) for a 6-h interval (from 8:00 till 14:00 UTC) obtained for the system almanac on March 2, 2015.

The results of statistical processing of the data from Fig. 5.7 are shown in Table 5.6.

The analysis of the data presented in Table 5.6 shows that not only VDOP values but also its variations are bigger that those of HDOP.

Table 5.6 Results of data processing

Parameter	Irkutsk	Moscow
$HDOP_{min}$	0.75	0.81
$HDOP_{max}$	1.35	1.22
$HDOP_{avg}$	0.989	0.959
Mean square deviation HDOP	0.117	0.113
$VDOP_{min}$	1.02	1.01
$VDOP_{max}$	1.87	1.82
$VDOP_{avg}$	1.423	1.438
Mean square deviation VDOP	0.213	0.228

Assuming the error of GLONASS pseudo-range measurement (see Table 5.4) equal to 4.5–9.2 m, the accuracy of altitude determination is, on average, 6.5–13 m which is insufficient, for example, for an SNS categorized approach.

Different methods and technical facilities are, therefore, developed and used to increase the accuracy of determination of civil user navigation parameters by SNS data.

5.3 Methods of Increasing the Accuracy of Navigational Sightings by SNS Data

5.3.1 Operation in a Combined Mode

One of the obvious methods to increase accuracy of navigational sightings by SNS data is to reduce GDOP. This can be done by increasing the number of satellites in the orbit group.

The enlargement of the orbit group of one system (GLONASS or GPS) is, however, economically inefficient. Solution to the problem was found in simultaneous operation of receivers in both systems (combined operation mode). In so doing, up to 25 satellites of both systems can be within the visibility zone, thus improving GDOP.

Figure 5.8 demonstrates HDOP and VDOP variations while operating in separate systems and in the combined mode for an airdrome located in the high latitudes (Igarka, Russia).

The results of statistical processing of the data from Fig. 5.8 are shown in Table 5.7.

As follows from Table 5.7, operation of a receiver in the combined mode allows the average value of VDOP and PDOP to be reduced by 30–60%.

But the modern state of GLONASS (quality of ephemeric and time-and-frequency provision, features of used signals) ensures a bit worse accuracy of navigational sightings than GPS. As a result, actual accuracy of determination of navigation

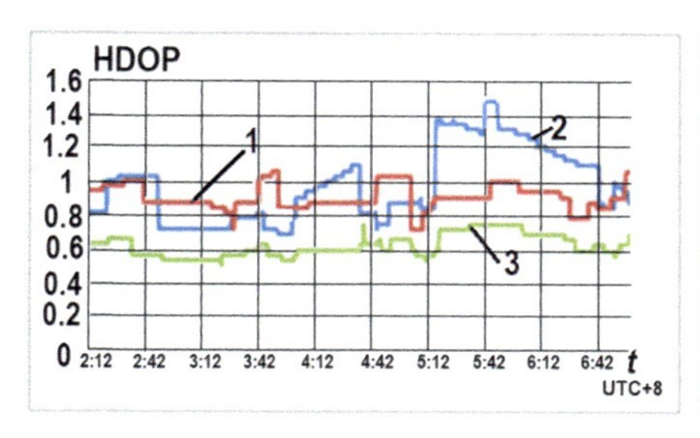

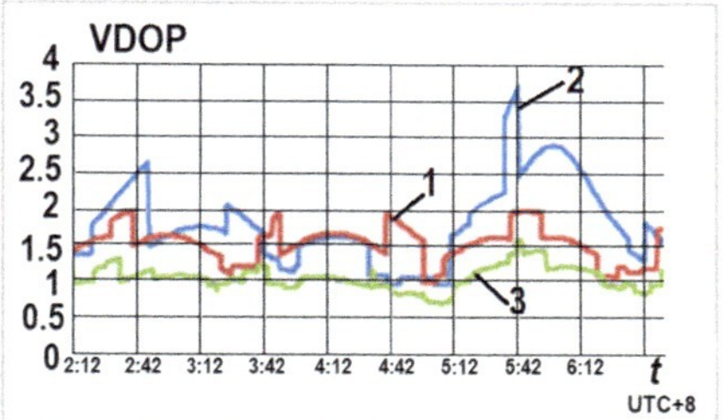

Fig. 5.8 Graphs of HDOP and VDOP variations: 1—operation using GLONASS; 2—GPS; and 3—GLONASS/GPS

Table 5.7 Results of data processing

Operation mode	$HDOP_{min}$	$HDOP_{max}$	$HDOP_{avg}$	$VDOP_{min}$	$VDOP_{max}$	$VDOP_{avg}$
GLONASS	0.75	1.1	0.906	1.0	2.0	1.512
GPS	0.75	1.5	0.978	1.0	3.7	1.793
GLONASS/GPS	0.5	0.75	0.621	0.7	1.5	1.052

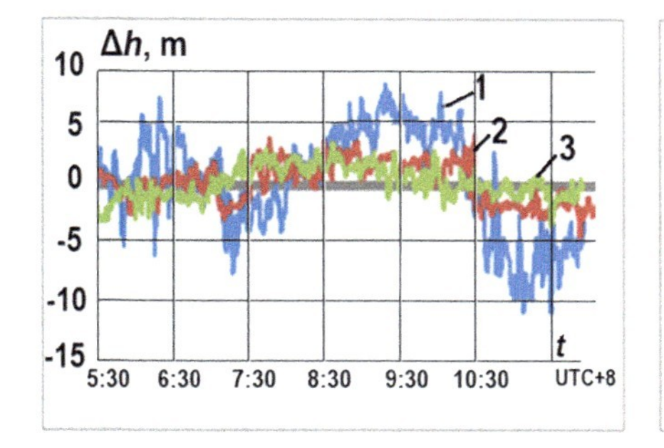

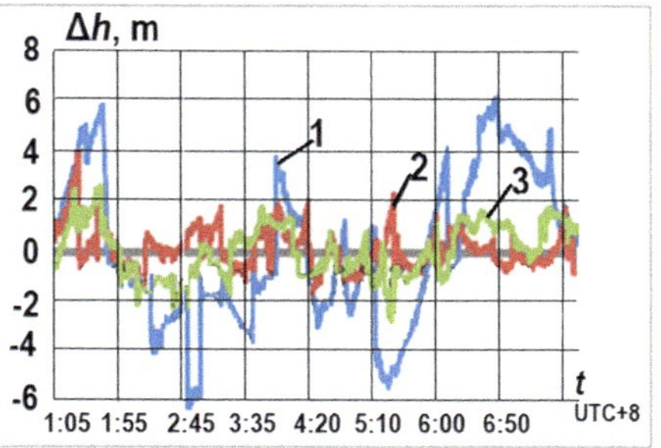

Fig. 5.9 Instantaneous error of altitude measurement: **a** natural experiment; **b** seminatural experiment

parameters in the combined mode may be even a bit worse than using only GPS but still better than using only GLONASS.

This was proved by results of natural (work with use of the real systems) and seminatural (work with use of signals of SN-3803M simulator) experiments carried out with SN-4312 receiver. Figure 5.9 presents graphs of variations of altitude measurement error (1—work with use of GLONASS, 2—work with use of GPS, and 3—work in the combined mode).

Calculated for the whole sampling of measurements during a seminatural experiment, the standard error of altitude measurement is 3.94 m (GLONASS), 0.53 m (GPS), and 1.19 m (GLONASS/GPS) for the given time interval.

So, nowadays it is appropriate to apply the combined mode if there are significant faults in GPS operation.

It should be also noted that sophistication of the design and software of combined receivers leads to increase in their cost.

5.3.2 Operation in a Differential Mode

The differential mode is based on the fact that the errors of pseudo-range measurements in two points of the earth surface contain correlated constituents for navigational sightings by the same satellites. Correlation is caused by systematic and slowly changing errors: ephemeris errors, satellite clock drift and additional ionospheric, and tropospheric delay of signals when propagating. Degree of correlation depends on the distance between reception points and is rather strong for SNS at the distance of 200–250 km. It may be assumed that the signal from a satellite to the points travels along a common radio path in the ionosphere and troposphere. If the measurement error is extracted at a reference station (RS), its subtraction from the measurements at the other RS (user equipment) causes the compensation of the correlated constituent.

To extract the measurement error, the RS antenna coordinates (reference coordinates) must be known exactly. For this purpose, its geodesic coupling on the ground is performed.

Comparing the reference coordinates and RS coordinates measured when receiving real signals from satellites, you can calculate corrections to measurements of either pseudo-ranges or coordinates. If in solving a navigational problem the user and RS work by the same satellites, the correction methods for coordinates and measured pseudo-ranges give the same result.

The calculated corrections are sent to the users in the RS area through a special radio link.

Implementation of the differential mode and, at the same time, external control of SNS integrality is accomplished by functional add-ons—differential subsystems.

A generic diagram of a differential subsystem is shown in Fig. 5.10. The most important role in the subsystem belongs to a datalink which determines its appearance, tactical, and technical capabilities. Two other components of a differential subsystem are RS and points of collecting and processing information which form corrections for users. To receive and use these corrections, a user must have corresponding receiving equipment, while an SNS receiver must have soft hardware necessary for correcting.

Depending on the coverage of the earth surface, there exists:

wide area differential subsystem;
regional differential subsystem;
local area differential subsystem.

According to the radio link used for transmission of differential corrections, ***Satellite-based Augmentation System*** (***SBAS***) and ***Ground-based Augmentation System*** (***GBAS***) are distinguished.

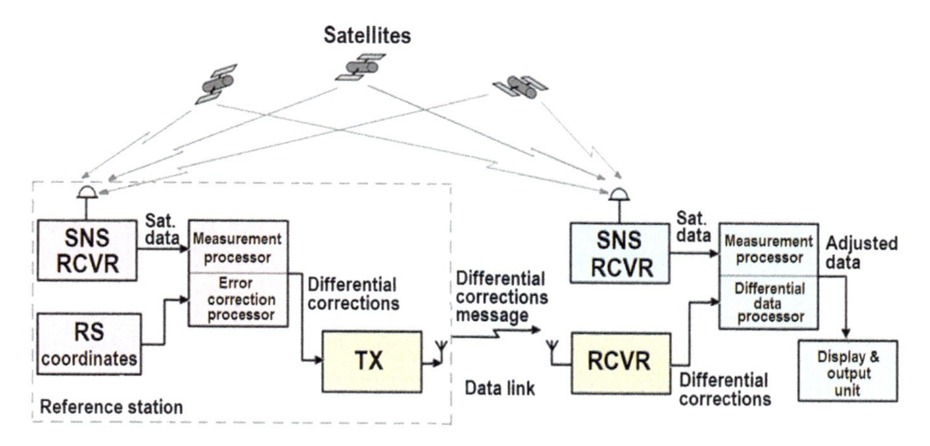

Fig. 5.10 Generic diagram of a differential subsystem

5.3.3 Functional Augmentations Recommended by ICAO

Wide area differential subsystems are designed to form corrections to the parameters measured by SNS on a significant part of the earth surface (e.g., within the whole continent). They include:

Wide Area Augmentation System (*WAAS*)—serves the territory of the USA including the state of Alaska;
European Geostationary Navigation Overlay Service (*EGNOS*)—serves the territory of Europe;
GPS-Aided Geo Augmented Navigation (*GAGAN*)—serves the territory of the Indian subcontinent;
Multifunctional Satellite-Based Augmentation System (*MSAS*)—serves the territory of the Far East;
Sistema Differencial'noj Korrekcii i Monitoringa (*SDKM*)—the system of differential correction and monitoring serves the territory of Russia. SDKM contains 22 RS and 3 GEO satellites of *Luch* type. It provides control over the navigation field integrality, formation of correcting information for GLONASS and GPS, and eventually Galileo as well.
BeiDou—serves the territory of China.

The system coverage is defined by the territory of RS location (from 8 for MSAS to 70 eventually for WAAS) and by coverage of correction datalink system which uses satellite communication channels. For transmission (retranslation) of corrections formed by ground-based master stations, geostationary satellites (from 2 for MSAS to 8 eventually for WAAS) are used. All wide area systems are, therefore, SBAS, while an operating range provided by a GEO satellite is 5,000–6,000 km. For effective correction of correlated constituents of the errors, the distance between RS should not exceed 400–500 km.

GEO satellites can also perform functions of additional navigation satellites radiating signals, for example, in GPS format.

To receive and extract correcting information, a user must have an individual receiving device.

Within SBAS coverage, aircrafts fly en route, in the terminal area, perform approach with APV-I and APV-II levels (approach procedure with vertical guidance, not meeting precision approach). Meanwhile, APV-I and APV-II approaches can be accomplished only in specific, assigned areas but not through the whole SBAS coverage area.

Regional differential subsystems or Ground-based Augmentation Systems (GBAS) are presented by two system types: ***Ground-based Regional Augmentation System*** (***GRAS***) and ***Local Area Augmentation System*** (***LAAS***).

GRAS is designed for navigational support of individual regions of a continent, sea, or ocean. Operating diameter can be from 400–500 to 2,000–2,500 km and is defined by the coverage of the datalink system. Unlike SBAS, users receive correcting information directly from a ground-based transmitter located at the distance of up to 1,000–2,000 km. GRAS includes Starfix, SkyFix, and Eurofix systems.

One of the main GRAS problems is necessity to transmit digital information (corrections) over long distances. For this purpose, datalinks with corresponding capacity are needed.

One of these links was created on the base of LORAN-C system [a modernized version is eLORAN (Enhanced LORAN)]. For data transmission, time-shifted pulses in a 8-pulse signal of standard LORAN-C are used.

GRAS can include one or several RS, a master station as well as means for transmission of correcting information and integrality signals to users. The information is formed at the master station or reference stations.

In LAAS, differential corrections are transmitted by a reference station located at a relatively small distance (up to 200–250 km) from the user receiver. Hence, these systems can provide a higher accuracy of position fix than WAAS or GRAS at the expense of a higher correlation of errors at the reference station and the user. Therefore, in aviation LAAS is applied at the flight stages where a higher accuracy of aircraft position fix is required—during approach and landing.

As a rule, the reference station is located directly at the site of airdrome RW. Coordinates of the reference station are defined with high (centimetric) accuracy. The reference station receives navigation signals from satellites, defines differential corrections, and then sends them to the aircraft receiver via an individual VHF radio channel of 108–117.975 MHz band. Along with the corrections, information about the system integrality and other procedure messages are transmitted.

So far, several types of aircraft LAAS landing systems have been developed (***GBAS Landing System***, ***GLS***). An example of such system is LKKS-A-2000 (Russia). In comparison with ILS, these systems have the following advantages:

small in nomenclature, easier in deployment and operation equipment which can be used for local airdromes;

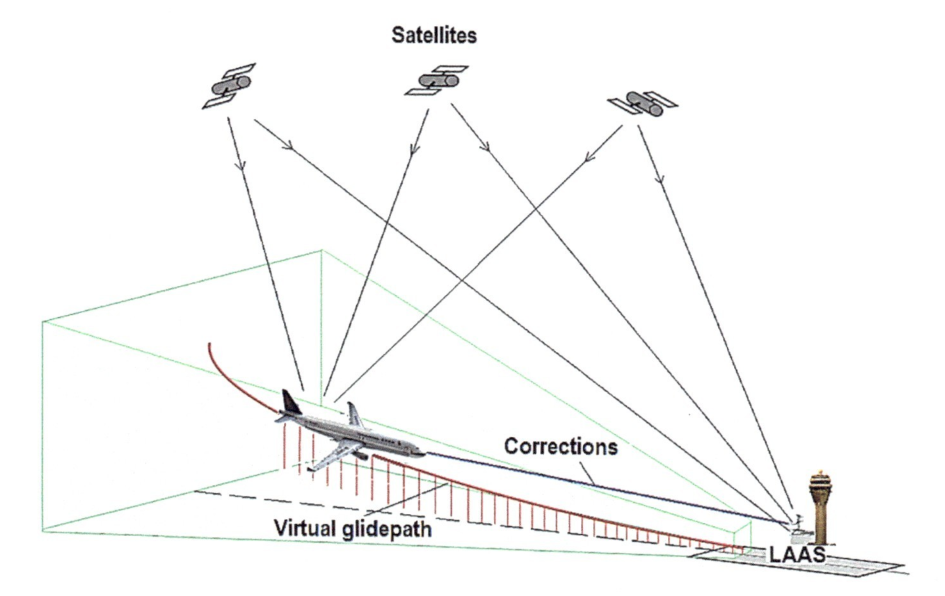

Fig. 5.11 LAAS landing

circular service area. A LAAS can serve approaches on all runways of the airdrome with both landing courses and under certain conditions—at several airdromes located nearby.
ability to implement optimal (and not only straight-line) approach trajectories from large distances. This enables to increase the airdrome capacity and flight safety in its area, to minimize adverse impacts of aircraft on the environment.
no effect of meteorological factors on glide path stability.

The approach trajectory (a virtual glide path) is calculated in the aircraft FMS computer using the information from the airborne database. Deviation of aircraft from the desired flight path is determined by comparing its actual (corrected by LAAS data) position with the calculated virtual glide path. Information about the deviations is supplied to an indicating device similar to the indicators used during ILS landing (Fig. 5.11).

Nowadays GLS ensure accuracy meeting CAT I. Eventually, a problem of ensuring more complicated categories will be solved.

5.3.4 Application of Pseudo-satellites

SNS augmentations in the form of ***pseudo-satellites or pseudo-lites*** (***PL***) can be used to increase the accuracy of aircraft positioning. A PL is a ground-based or NES

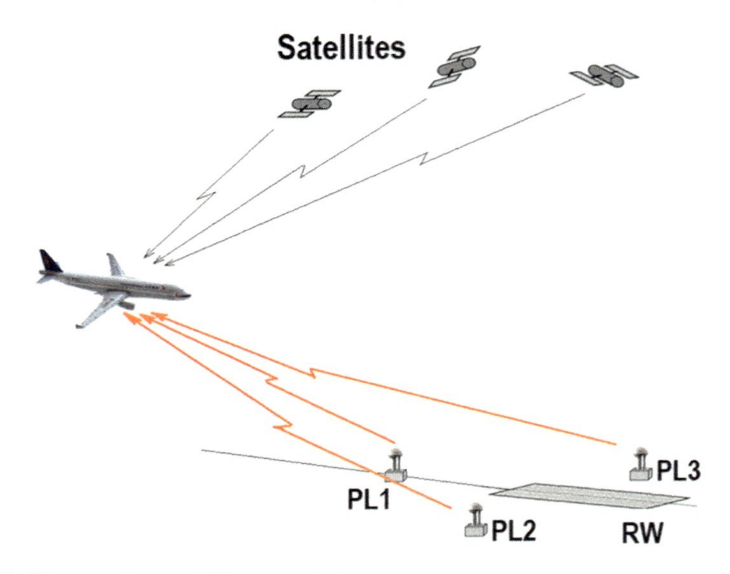

Fig. 5.12 PL network as an SNS augmentation

(near-earth space)-based radio transmitting device which emits signals synchronized with SNS signals, and the parameters and the format of the signals are close to those of the SNS signals (Fig. 5.12). The PL coordinates must be known precisely.

There are a number of reasons for using PL, primarily:

increasing navigation availability in degraded environments;
increasing integrity for safety-of-life applications;
increasing accuracy by improving the overall geometry, particularly in the vertical dimension (VDOP).

Precise landing, emergency services in difficult environments and precise positioning and machine control are few examples where PL technology can be employed.

The use of PL ensures DOP enhancement due to increase in the number of navigation information sources provided they are placed in the terminal area rationally. In practice, therefore, a problem arises of determining a sufficient number of PL and their optimal placement to ensure the required accuracy of aircraft position fixing.

For example, let us assume that two PLs are installed in the runway area at a distance of 1 km from its threshold and with a displacement of 1 km from its axis. Consider the variation of a mean VDOP in the cases (Fig. 5.13a):

(a) PL1 is shifted in parallel to the RW axis at a spacing of 1 km, and PL2 is located at the reference point;
(b) PL1 is located at the reference point, and PL2 is shifted perpendicularly to the RW axis at a spacing of 1 km.

To explore the VDOP variations, a model of orbital motion and data of the real GLONASS almanac were used. The analysis was performed at a real airdrome

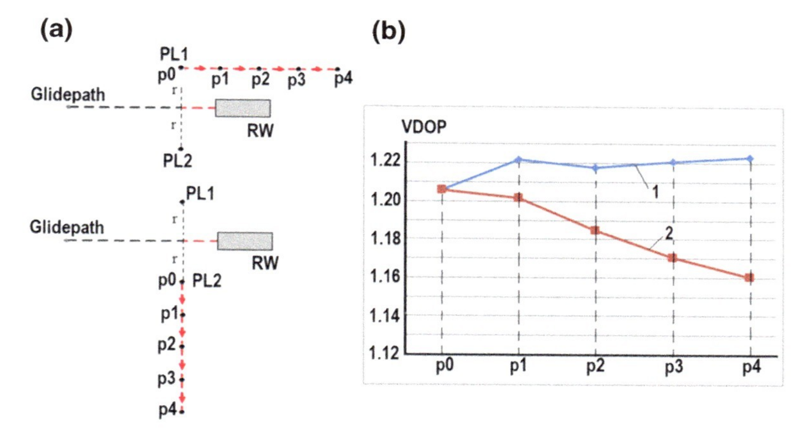

Fig. 5.13 VDOP in different PL configuration: **a** ways of PL shift; **b** VDOP variation

(Russia, high latitudes). The research results are shown in Fig. 5.13b (curve 1—with PL1 shifted; curve 2—with PL2 shifted).

As can be seen from the presented results, VDOP and therefore accuracy of determining the altitude along the landing trajectory depend on the PL position relative to the RW. Thus, the problem of determining the optimal position of PL is urgent. For its solution, it is helpful to use methods of optimization theory and to choose, as an optimality criterion, the minimum of the VDOP mean value either at a specific point or along the whole landing trajectory.

The problem solution is complicated by the fact that satellites are continuously moving along their orbits. Hence, there is no single optimal position of PL which can provide the minimal VDOP. Indeed, if we solve the problem of determining the optimal position for one PL with respect to GLONASS system with 1 h discretization, we will obtain 192 solutions for a complete period of orbital motion repetition (8 days).

Apply the Nelder–Mead method as a method of determining the PL optimal position. In doing so, we will limit the zone of permissible positions of PL to an area of 50 × 50 km in latitude and longitude with the center coinciding with the RW center.

The results of solving the problem of determining the optimal position of one PL with 1 h discretization are shown in Fig. 5.14.

The received results show that a considerable part of the found optimal positions of PL is grouped close to the landing trajectory.

To find a single solution, we will construct a histogram of distributing the found coordinates of PL in latitude and longitude (Fig. 5.15a, b correspondingly) and choose those which have the maximal number of hits within the provided intervals of 10 m in latitude and longitude. The PL position found in this way is, in fact, quasi-optimal.

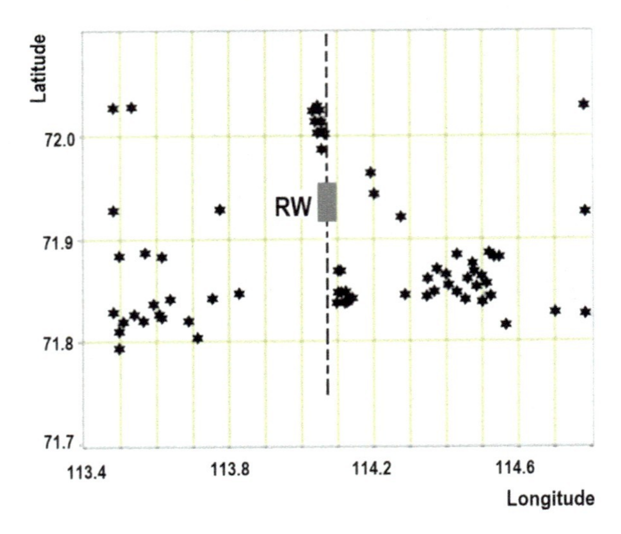

Fig. 5.14 Points of PL optimal position

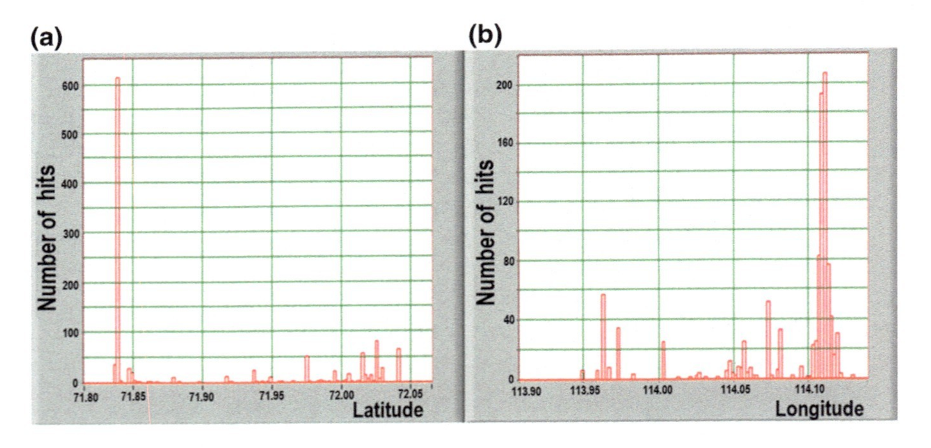

Fig. 5.15 Histograms of distributing the PL coordinates in latitude and longitude

Figure 5.16 demonstrates a graph of day variation of VDOP without using PL (curve 1), using PL located at the Middle Marker point (curve 2) and using PL located at the quasi-optimal point (curve 3).

Mean values of VDOP are: without PL—1.51; with PL located at the Middle Marker point—1.17; with PL located at the averaged quasi-optimal point—1.11; and with PL repositioned to an optimal point hourly—1.07.

Thus, the use of one PL enhances the average VDOP in the terminal area by 22.5%, while optimization of the PL placement ensures an additional enhancement of VDOP by 4–7% (for the given example).

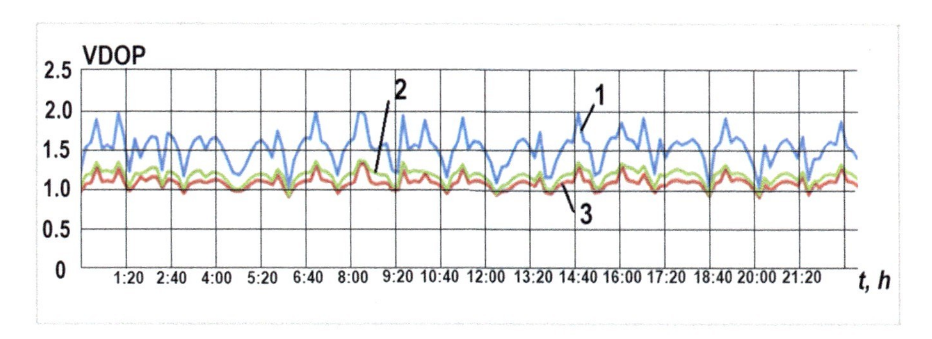

Fig. 5.16 Day variation of VDOP

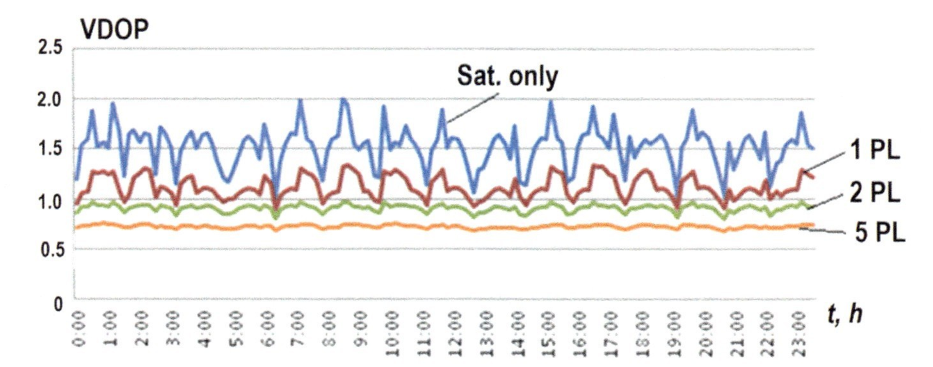

Fig. 5.17 Day variation of VDOP with a different number of PL

Time realizations of the VDOP mean value for a 24-h interval with a different number of optimally placed PL are shown in Fig. 5.17. The received results show that an increase in PL number leads to decreasing the VDOP and its variations range. The efficiency of PL application, however, drops with the increase in their number, and using more than three PL is unreasonable for increasing the accuracy of position fixing.

It should be also noted that the efficiency of PL application increases with DOP degradation, for example, when the number of satellites in the radio-visibility zone decreases, their signals get shadowed by airdrome objects, terrain relief is rough, interference environment is complex.

PL use has some issues:

the GPS и GLONASS L1 band is reserved for aeronautical use, to use PL transmitting in this band for other uses may require a change in legislation.

interference with existing GNSS signals: The issue is to quantify and minimize this interference;

the "near/far" problem: Received PL signal powers vary greatly over their operational range and can be significantly stronger than satellite signals. Receivers therefore

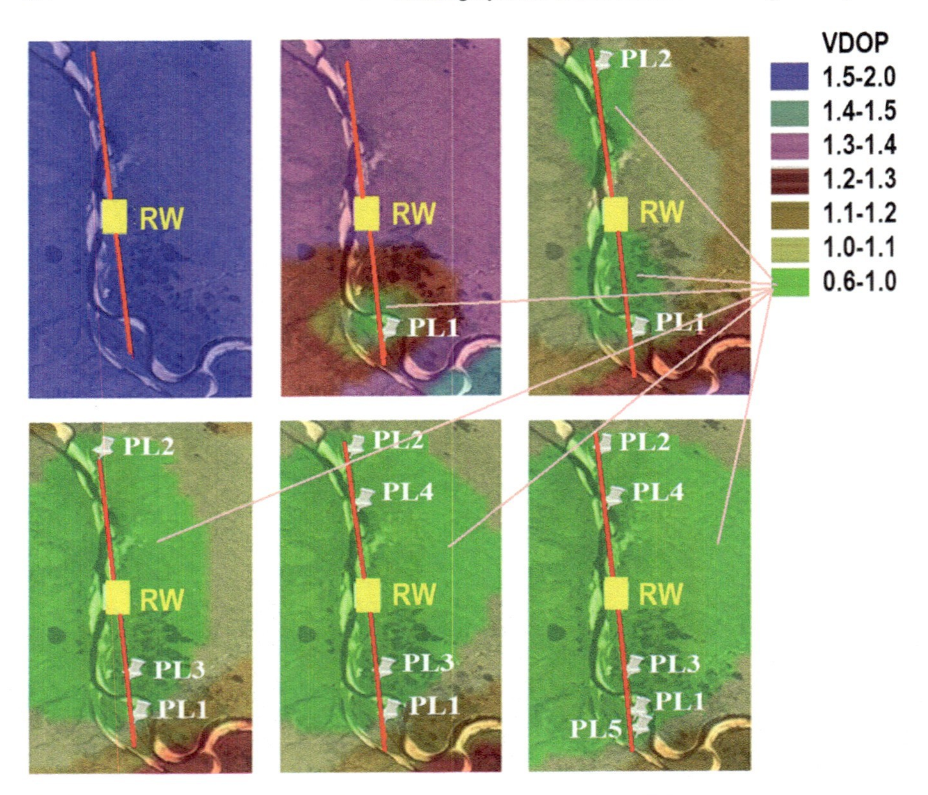

Fig. 5.18 Areas of equal values of VDOP with a different number of PL

require a large dynamic range, much larger than standard GNSS receivers unless specific counter measures are employed in the transmitter.
synchronization between the pseudo-lite clocks;
monitoring of PL performance and their signals.

Moreover, there exists a problem of dependence of an area size with the minimal VDOP on the flight altitude and the number of PL used.

To illustrate this problem, we built fields of equal values of VDOP at the height of 600 m above the ground for the GLONASS system augmented with a different number of PL (Fig. 5.18). The received results show that an increase in number of PL leads to increase in the area with the best VDOP. When using three optimally placed PL, this area includes the whole landing trajectory and the approach sector.

Figure 5.19 shows the areas of the best values of VDOP received for different altitudes (with three PL used). The received results show that the area of the best values of VDOP decreases as the flight altitude falls, which may cause problems at the stage of final approach.

Figure 5.20 demonstrates distribution of the areas with equal values of VDOP in vertical plane passing through the RW axis with different number of PL used. It is

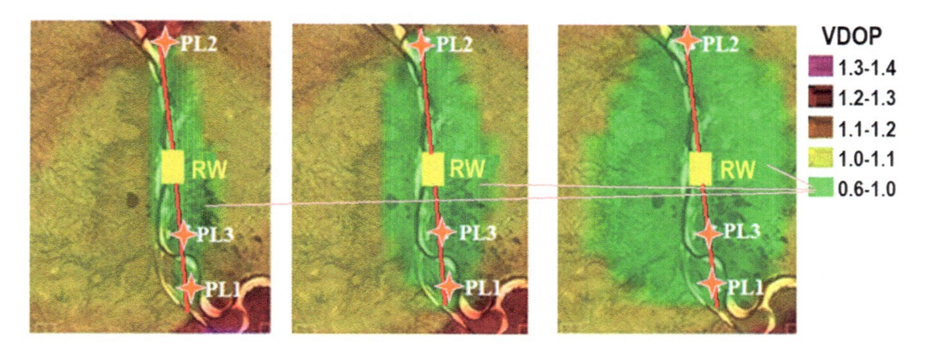

Fig. 5.19 Areas of equal values of VDOP at different altitudes: **a** $H = 30$ m, **b** $H = 300$ m, and **c** $H = 700$ m

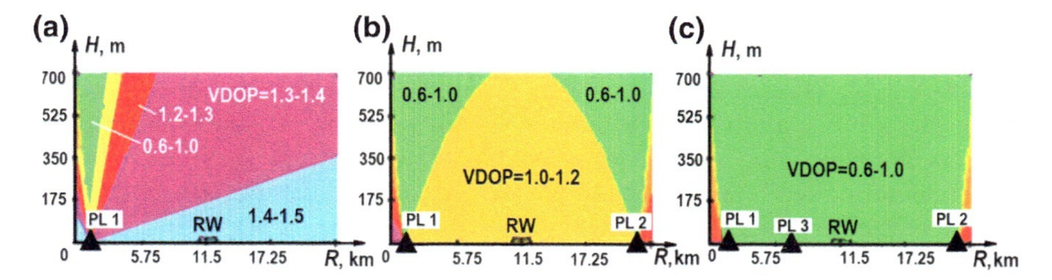

Fig. 5.20 Areas of equal values of VDOP in vertical plane

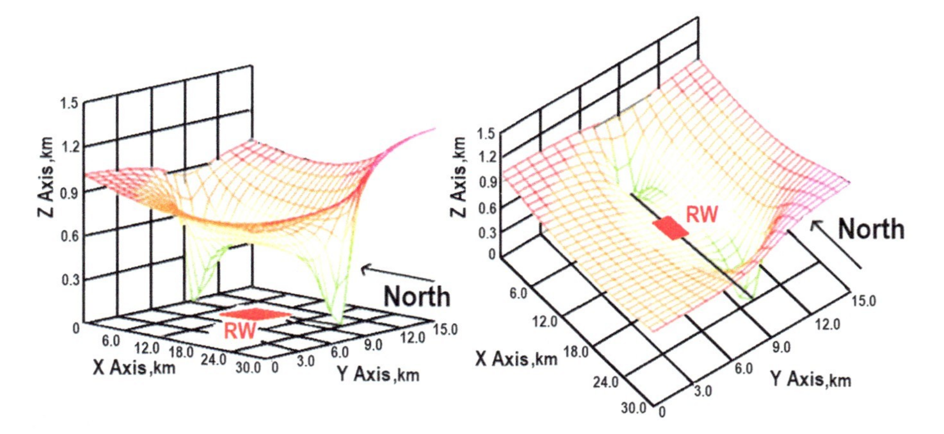

Fig. 5.21 Spatial distribution of VDOP values with two PL used

seen that with one PL used, the best VDOP = 0.6–1.0 is present when flying over the PL and it degrades to 1.4–1.5 as the aircraft approaches the RW (Fig. 5.20a). With three PL used as Fig. 5.19 shows, the area of the best values of VDOP remains constant along the entire glide path (Fig. 5.20c).

Figure 5.21 demonstrates spatial distribution of VDOP values with two PL used.

The received results show that it is helpful to use three PL optimally placed in the RW area to achieve the highest accuracy of aircraft position fixing during approach and landing. One PL used provides the best values of VDOP only within the cone-shaped area with the vertex coinciding with the PL.

5.4 Design and General Performances of LAAS of LKKS-A-2000-Type

LKKS-A-2000 (*Lokal'naja Kontrol'no-Korrektirujushhaja Stancija, Russia*) is designed to form and transmit the correction and operation information in real time to users through a radio link as well as to ATM ground services and GLONASS and GPS monitoring services through wired data lines and satellite radio links.

The transmitted data include [6]:

differential corrections to pseudo-ranges and the speed of pseudo-range changes;
information about the current and projected state of GLONASS and GPS orbital groups within the coverage area of the station;
information about permissibility of performing routine operations by an aircraft with the use of GNSS data in the independent and differential operation modes of the airborne receiver with account of requirements for accuracy, integrity, availability, and continuity of a satellite navigation signal determined by RNP for a specific routine operation;
integrity control data (numbers of rejected satellites) when receiving signals from navigation satellites of GLONASS and GPS staying within the visibility area of the station;
data of quality control for a navigation satellite signal (SQM), prediction of GLONASS, and GPS integrity level within the coverage area of the station as well as additional navigation parameters in SARPS ICAO format at GBAS and performance characteristics of the station.

Differential data and other information are transmitted in the frequency range of 108–117.975 MHz (VDB) and 118–135.975 MHz (VDL-4).

A block diagram of the station is shown in Fig. 5.22.

The station can operate in the independent (monitoring and output of data about the state of orbital groups to users), differential (output of data about pseudo-range corrections and on-line monitoring) modes as well as in independent and differential modes jointly.

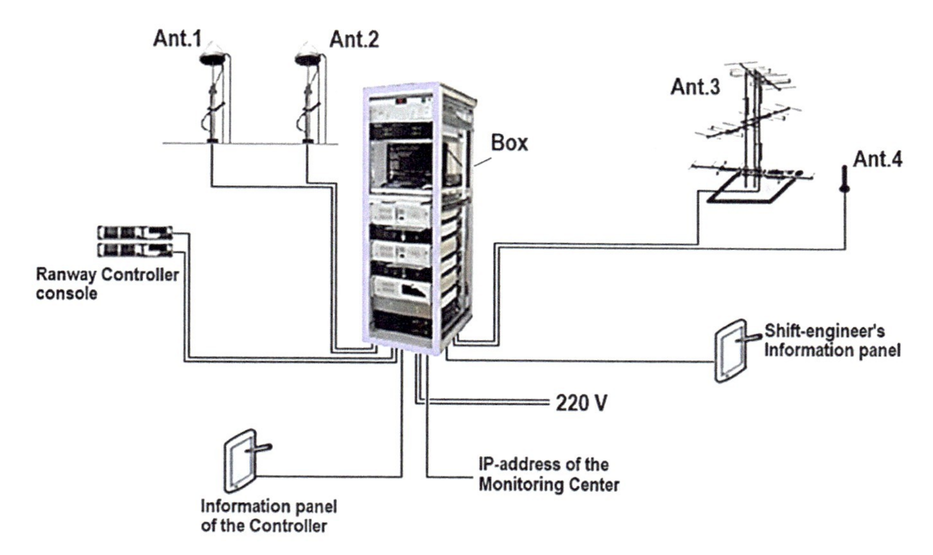

Fig. 5.22 Block diagram of the station

The station includes the box antenna-feeder devices of satellite receivers and data transmission lines, a controller's and a shift-engineer's consoles, and a controller's and a shift-engineer's computers.

The general performances of the station are given in Table 5.8.

A functional block diagram of the station is shown in Fig. 5.23.

Signals from GNSS satellites within the coverage area of the station enter all inputs of the satellite receivers through antennas and a power divider.

In UCSD, a complex analysis of pseudo-range corrections calculated by the receivers and the speed of their change is performed through programs of monitoring, navigation information quality control and integrity. Assessments obtained for each navigation satellite form the correction information for users.

The software of integrity control ensures detection of parameters exceeding threshold values with a probability of 0.98 in under 3 s. The correction information for this satellite is marked with a satellite rejection sign according to SARPs standard. If the number of non-rejected satellites is less than 4, the station generates warning about uncertainty of the information transmitted.

There is control of output correction information in the system. The satellite receiver of the output control system working in the differential mode determines its coordinates using the corrections calculated by the station.

The obtained values of the corrected coordinates are compared to the true (geodesically precise) coordinates of the antenna of the station.

If the difference between obtained and true coordinates exceeds the provided threshold value (10 m for any parameter—latitude, longitude, altitude), the station

Table 5.8 General performances of LKKS-A-2000

Item	Normative standard
Used GNSS systems	GLONASS, GPS
Period of data update and output, s:	
Differential data	0.5
Data of the reference station	1.0
LKKS identifier	15.0
FAS unit	15.0
Prediction of satellite availability (readiness)	15.0
Operational frequency of data transmission through a radio link, MHz	108.00–117.995
VDB transmitter power, W	50
Time of readiness for operation, s	<160
Coverage area for landing:	
in horizontal plane, km	≥37
in vertical plane, deg.	≥7.0
Coverage area for navigation	Radio visibility
Integrity with SQM function	$1–2 \times 10^{-7}$
Warning time, sec	<6
Continuity	$1–8 \times 10^{-6}$ any 15 s
Availability	0.99–0.9999

regards the situation as a failure and forms a warning about uncertainty of the transmitted correcting information.

The information formed by LKKS-A-2000 is presented to users through a VHF transmitter of one of the station sets, the transmitting antenna (Ant. 3) being connected to its output through the electronic switchboard. The transmitter of the other set is connected (on "hot" standby), but the voltage of +28 V is not supplied to its output cascade.

The channel for transmitting the correction information uses TDMA method with the fixed structure of frames (frame duration is 500 ms), and modulation type is D8PSK. The transmitter power is 50 W.

The system implements the program control of integrity, continuity, and certainty of information supplied by each VHF transmitter. This information is received by the receiving antenna (Ant. 4), enters each VHF receiver and, after decoding, is compared to the transmitted information.

The UCSD contain industrial computers performing calculative-switching operations. The link between UCSD1 and UCSD2 computers is provided through Hub 1 и Hub 2 switchboards.

Data transmission from UCSD to remote consoles and the central computer of ATC controllers and engineering staff of the airport is performed in RS232, Ethernet, and other agreed standards through wired and satellite data transmission lines.

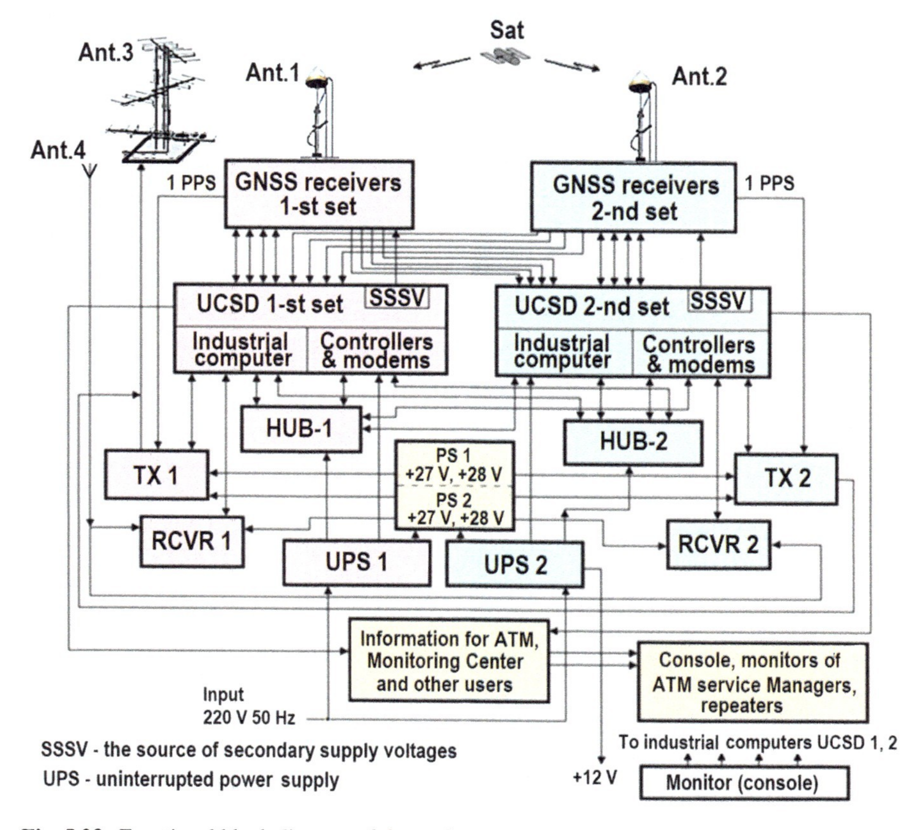

Fig. 5.23 Functional block diagram of the station

While distorting in the data transmission lines, the station generates a warning about the uncertainty of the information transmitted.

If both the main and reserve LKKS-A-2000 sets fail, an arithmetic mean differential correction for each satellite is calculated and transmitted.

Each UCSD computing and switching device of either set forms information about predicted availability of satellite groups in the station coverage area and about the performance status of LKKS-A-2000 modes and units and supplies this information to the datalink or to the remote consoles.

5.5 Summary

Satellite navigation is one of the fastest growing and widely used technologies. It rapidly encloses new spheres of humans' areas of activities. Perhaps, the appearance of satellite navigation systems and its usage in non-military area is one of the most important achievements of humanity in the end of XX century.

Modern satellite navigation systems which are described in Chap. 5 have a number of technical characteristics which put these systems on special, leading place in the row with the other technical navigation means. These characteristics are mainly high accuracy (first dozen and in specific conditions, entity of meters) and globality of operating area.

Here, we have to specify that GNSS is a complicated complex of technical means (ground complex of control, satellite's orbit groups, vast park of apparatus of consumers of different purpose, and different functional possibilities). Deployment of such systems and their operation provisioning requires serious financial expenses, and however, the economical effect they have exceeds the costs. This serves as a great motivation for further GNSS development, also on behalf of the civil aviation.

GNSS is the basis of navigation concept realization which is based on the characteristics (PBN). Provision of new operational benefits from the perspective of flights safety and communication means efficiency, navigation, and surveillance (CNS) which based on the wider usage of 4D approach by instruments according to the global plan of ICAO in the area of PBN is based on wider usage of GNSS.

Thus, in midterm perspective safer and more effective flights operations, getting additional operational benefits will become possible in case of implementing the new technologies of dual-frequency (DFMC) GNSS via a number of satellites constellation.

That is why it is logical that aeronautical navigation analyses GNSS usage possibilities and different methods of using GNSS to perform landing are examined. This will allow to perform approach using optimal and, possibly, curved trajectories and will solve some problems common for ILS.

However, some problems of using GLS for landing with accuracy corresponding with the high categories of landing meteorological minima still exist. Section 5.3 shows the results of conducted under author's supervision work on usage of pseudo-lites as functional addition to GLONASS. These results show the efficiency of pseudo-lites usage in the landing phase.

5.6 Further Reading

Satellite navigation systems such as GPS and GLONASS as well as common principles of satellite navigation are described in many books of various authors.

The classic more detailed textbook on GLONASS was published in Russia in 1979 by V. Kharisov and A. Perov, and its latest edition is [1].

Information on construction and operating principles, main technical characteristics of Russian-made aviation receivers, can be found in [2].

Official information on declared characteristics of GPS and GLONASS systems, their content and orbital groupings' parameters as well as the parameters of free-access signals and algorithms of their long-range system's codes formation can be found in [3, 4].

There is a wealth of information on GPS at http://www.gps.gov and https://en.wikipedia.org/wiki/Global_Positioning_System. Here, you can also find links to the resources on this topic.

There is a wealth of information on GLONASS at https://www.glonass-iac.ru and https://en.wikipedia.org/wiki/GLONASS.

Information about research of the integrated accuracy field GLONASS in the optimal placement of pseudo-lites in the aerodrome zone can be found in [5].

References

1. Perova AI, Kharisova VN (eds) (2010) GLONASS. Principy postroeniya i funkcionirovaniya. [GLONASS. Principles of construction and operation], 4-th edn. Radiotehnika, Moscow (in Russian)
2. Skrypnik ON (2014) Radionavigacionnye sistemy vozdushnyh sudov [Radionavigation systems of aircrafts]. INFRA-M, Moscow (2014) (in Russian)
3. Global Positioning Systems Directorate Systems Engineering & Integration Interface Specification IS-GPS-200 (2018) NAVSTAR GPS space segment/navigation user segment interfaces. Download from https://www.gps.gov/technical/icwg/IS-GPS-200J.pdf
4. Interface Control Document (2008) Navigational radio signal in bands L1, L2, 5.1 edn. Moscow. Download from http://russianspacesystems.ru/wp-content/uploads/2016/08/ICD_GLONASS_eng_v5.1.pdf
5. Skrypnik ON, Kargapol'cev SK, Sizykh VN, Daneev AV, Aref'ev RO (2018) Characteristics of the integrated GLONASS accuracy field in the optimal placement of pseudo satellites in the aerodrome zone. Adv Appl Discrete Math 19(2)
6. Lokal'naya kontrol'no-korrektiruyushchaya stanciya LKKS-A-2000. Rukovodstvo po tekhnicheskoj ehkspluatacii [Local monitoring and correcting station LKKS-A-2000. Technical operation manual]. 3-rd edn. LLC "NPPF Spektr", Moskow

Chapter 6
Multiposition (Multilateration) Surveillance Systems

Multilateration systems (MLAT) are designed for determining the coordinates of moving objects such as aircrafts or airfield's facilities, equipped with aids of impulse radio signals' emission. Fundamental operating principle of MLAT is the usage of multilateral principle which is also used in the time-difference long-distance radio navigation systems such as LORAN-C and Chayka. This principle is described in Sect. 1.7.

MLAT systems found its use relatively recently in the beginning of 2000s. Its appearance can be explained by the necessity of surveillance aids' efficiency increase under conditions when the field of observation, created by another technical aid (radio locators, direction finders) does not solve the tasks of air traffic management on the necessary modern level. Herewith, it turned out that MLAT systems are extremely efficient, especially in conditions of complicated terrain and can be used in the airfields' zones as well as on the air routes.

Despite of MLAT systems being first of all the surveillance aid, author considered possible its overview in this book, because systems execute the navigation principle of coordinate determination—hyperbole fixing.

In Sect. 6.1, common characteristic of MLAT systems is given; its purpose, content, aspects of use, and unified block diagram of the system is described.

In Sect. 6.2, principle of coordinates' determination in MLAT system with the help of mathematics is described.

In Sect. 6.3, typical variants of constructing MLAT systems and methods of synchronization of ground-based receivers of the systems are studied.

Section 6.4 studies factors that influence the positioning accuracy in the MLAT system and the variations of the system's operating area depending on the architecture of system and on the aircraft's flight height.

In Sect. 6.5, main requirements to the MLAT system characteristics which are installed by consumers are given, advantages and disadvantages of the system are described. Russian-made systems are also mentioned in this section.

O. N. Skrypnik, *Radio Navigation Systems for Airports and Airways*,
Springer Aerospace Technology, https://doi.org/10.1007/978-981-13-7201-8_6

6.1 General Characteristics of Multilateration Systems

Multilateration systems are designed to determine the position of moving objects (aircraft, airport vehicles) equipped with devices emitting radio pulse signals. The system operation is based on the principle of multilateration. This principle is used in Chayka (Russia) and LORAN-C (the USA), widely known pulse-phased differential distance long-range navigation systems.

The multiposition surveillance systems have been called multilateration system (MLAT) or wide area multilateration system (WAM). The WAM system provides surveillance in the en route airspace while MLAT is used to ensure surveillance of the terminal area airspace or ground traffic at the airport.

The advent of MLAT technologies was caused by necessity of increasing the efficiency of surveillance facilities when the field of vision created by other technical means could not provide solutions to ATM and ATC tasks at the required modern level. The issues that can be solved by means of MLAT include:

surveillance in the airspace where it is not currently carried out and implementation of traditional surveillance facilities—radar systems—is impossible for some commercial, technical, or environmental reasons;
surveillance in the areas where the ground profile makes it difficult to use radar sets because of blockage;
provision of a surveillance source as an augmentation of existing radar sets;
increase in safety of instrument flight rules in areas having dangerous ground profile;
cost savings as it eliminates costs of installation, maintenance, full-range operation, and enhancement of existing surveillance radar systems;
monitoring of altitude hold of aircraft approved for reduced vertical separation minimum (RVSM) operation.

So the MLAT system can be used as a fairly efficient additional or alternative source of surveillance data in order to ensure ATM including the functions of flight safety control which are currently performed by secondary surveillance radars (SSR).

A great advantage of multilateration systems is their ability to track not only aircraft but also airdrome vehicles on the surface of airdrome. MLAT application for airdrome traffic surveillance is possible if the vehicles are equipped with proper transponders capable of operation on the ground. Thus, Mode S transponders continue to transmit self-generated signals and receive selective interrogations when the aircraft is on the ground but Mode A/C transponders are prohibited to reply the interrogations when the aircraft is on land in order to reduce interference for nearby radar systems.

MLAT as a technical system is a means of secondary radio location where the aircraft coordinates are measured with differential distance method by using the stations located on the ground. The differential distance method of determining a position considered in Sect. 1.6 is also known as multilateration method which has been a reason why the system is called MLAT.

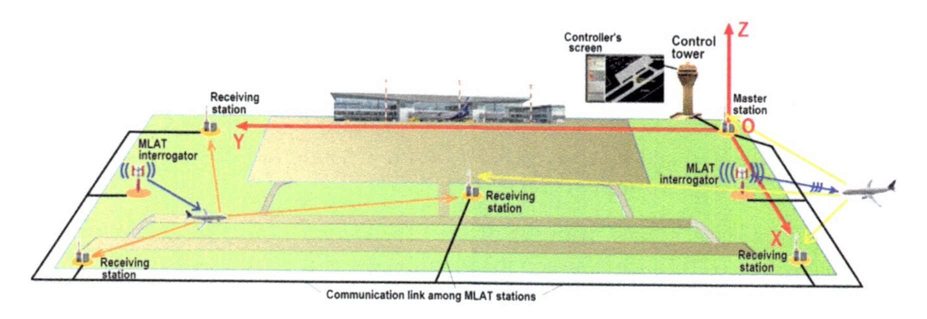

Fig. 6.1 Placement and interaction of the MLAT elements in the terminal area

But from the perspective of radio navigation, MLAT is an ***inverted differential distance system***. Unlike the conventional differential distance method, it uses one source of signal radiation (the aircraft transponder) and a network of spaced receivers.

The aircraft attitude is defined with positional method as a point of intersecting two or more surfaces of position having forms of hyperboloids—surfaces of equal range differences.

To implement the differential distance method, it is necessary to have, in the given coordinate system, a network of spaced ground stations with known coordinates (Fig. 6.1) and to measure time differences of arrival (TDOA) of the signals from onboard transponders. There is a master station within the network and all the other receiving stations fix the time of arrival of a signal from the onboard transponder relative to the time when the master station receives the same signal. Processing of time differences of arrival of signals at the station pairs helps to figure out range differences and to solve the task of calculating the aircraft coordinates. TDOA is processed in the MLAT controller which can be located at the master station or at the control tower.

With three stations available, you can determine the aircraft position on the plane (i.e., on the earth's surface) and with four or more stations, it is possible to determine unambiguously the aircraft position in space (3D navigation). It should be noted that for MLAT, the aircraft coordinates are determined on the ground; hence, the task solution does not depend on any onboard means of navigation—only the airborne SSR transponder is needed.

In addition to receiving stations, MLAT can include transmitting stations capable of interrogating the onboard transponders. It is necessary when within the system coverage there are no other interrogators with the signals causing formation of SSR reply signals. MLAT also has a principal possibility to use interrogation signals and subsequent reply signals from an aircraft for measuring the ranges to the aircraft supplementing differential distance measurements.

MLAT is a part of *advanced surface movement guidance and control systems* (ASMGCS systems) and is designed to monitor an aircraft during landing and taxiing at the airdrome, thus enhancing the efficiency and safety of the operations in conditions of high-density air and airdrome traffic. The WAM systems are in addition used for monitoring an aircraft en route and in the approach areas.

A modern MLAT receiver can process up to 30,000 reply signals per second. The TDOA measurement error is no more than 1 ns and it is equal to an instrumental error of range difference measurement of 0.3 m. The aircraft position accuracy is from 3 to 150 m with resolution ability:

for air routes—at least 60% with the distance between the aircraft of 2 nm and at least 98% with the distance between the aircraft of 4 nm;
in the terminal area—at least 60% with the distance between the aircraft of 0.6 nm and at least 98% with the distance between the aircraft of 2 nm.

The system provides surveillance of at least 500–600 aircraft with the transponder operating in Mode S and at least 100 aircraft in Mode A/C. Signal processing time including delays in communication channels does not exceed 250–500 ms. Periodicity of updating the information of aircraft coordinates is 1 s which is much more frequent than standard surveillance radar means provide (12 s or 4.7 s). This ensures higher accuracy of aircraft surveillance which is especially important in conditions of high density and intensity of air and airdrome traffic. The network of four airdrome stations has coverage of approximately 20 km.

MLAT can operate jointly with primary and secondary control radars, ADS-B terminals.

In practice, TDOA of reply signals is measured with inaccuracies. The increase in the number of ground stations above four will improve the accuracy of determining the aircraft position since the navigation task will be solved with measurement redundancy. The task of determining the position is, in this case, an optimization task which can be solved, for example, with the use of the least square method or extended Kalman filter.

As in any secondary radar, the MLAT ground-based interrogator emits an interrogation; the airborne transponder receives it and forms a reply which is received by MLAT ground stations. The reply signal contains information about the aircraft and its movement parameters, first of all its altitude and identification and besides it is used for determining the aircraft position. The 1,090 MHz self-generated signals of an airborne transponder can also be used as reply signals. If the reply signal contains information about the aircraft height, it is possible to determine its spatial attitude with only three MLAT ground-based receiving stations available.

Depending on the way how the airborne transponder forms the signals, the MLAT systems can be divided into active and passive ones. The passive system includes only receivers while the active one has one or several transmitting stations for interrogating the SSR airborne transponder. The major advantage of the active system is that it is independent on the other sources of initiating the data transfer from an aircraft. But its main drawback is that the signal of its interrogator presents interference for the SSR system itself on 1,030 and 1,090 MHz channels.

The MLAT ground stations receive aircraft signals from all directions and the aircraft coordinates are calculated according to the delay of *time of arrival (TOA)* of the signal at different stations. The measured delays between the times of arrival

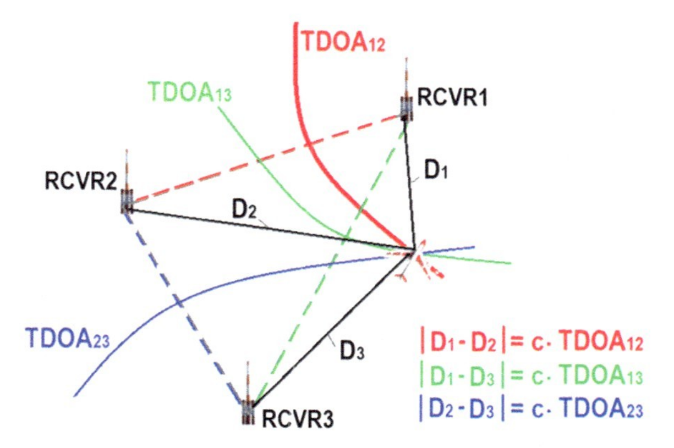

Fig. 6.2 Aircraft positioning on a plane

of signals at the station pairs allow the surfaces of position to be constructed. The surfaces of position are surfaces of equal range differences (hyperboloids) with the intersection point being the sought aircraft position. When intersecting with a plane, the hyperboloids make up hyperbolas (Fig. 6.2) with an intersection point allowing the aircraft coordinates to by spatially determined (2D navigation).

As mentioned above, for measuring TDOA and calculating the aircraft coordinates, all receiving stations must have synchronized time scales which are used to fix TOA of airborne transponder signals; or the measurements should be performed on a common reference scale of one of the stations called the master station.

Generally, the TOA (times of arrival of a reply signal from a particular aircraft) fixed by the ground receiving stations and master station are sent to the server (MLAT controller). Sometimes, the MLAT controller can be together on the master station. The MLAT controller calculates TDOA for each signal pair between the master and other stations. With four stations available, it is possible to get three values of TDOA which allows a navigation task of aircraft 3D positioning to be solved with the use of differential distance method.

The MLAT block diagram illustrating its composition and interaction of its functional elements is presented in Fig. 6.3.

It should be noted that the channels of data exchange between the stations and MLAT controller are of great importance for the MLAT system since, when the data are transmitted through the channels, additional signal delays occur. The system control principles meet modern requirements—there is both local and remote control and status monitoring.

It should be noted that for implementation of differential distance method, there is no need to know the time when the airborne transponder emits the signal.

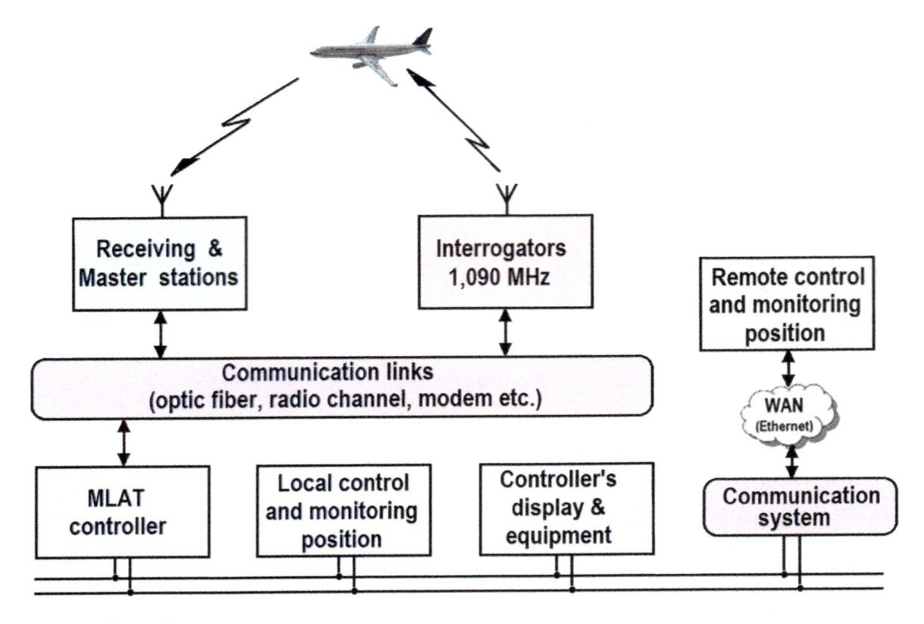

Fig. 6.3 MLAT block diagram

6.2 Principle of Aircraft Positioning in the MLAT System

Let us assume that the aircraft with the airborne transponder emitting a reply signal is in point *M* (Fig. 6.4). In a rectangular coordinate system, the aircraft position is characterized by vector

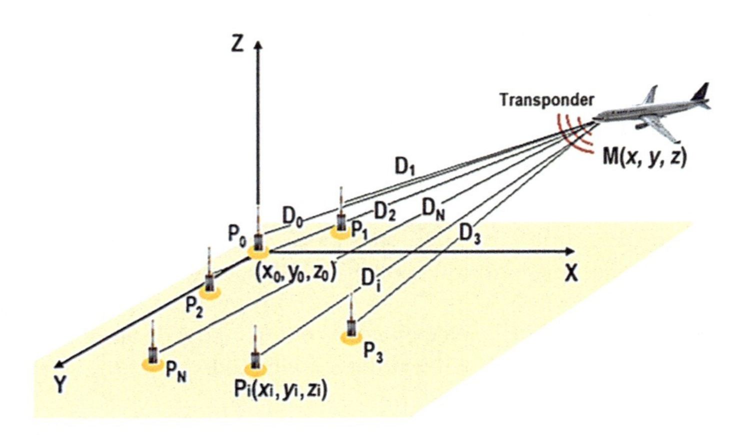

Fig. 6.4 Interaction of aircraft transponder with the MLAT stations

$$\mathbf{M} = \{x, y, z\}.$$

The aircraft reply signal is received by MLAT receivers N located in points P_0, P_1, ..., P_i, ..., P_{N-1} with known coordinates.

The position of a ith receiver (its antenna) is characterized by coordinate vector

$$\mathbf{P}_i = \{x_i, y_i, z_i\}, \quad 0 \leq i \leq N.$$

The distance from the transponder (the aircraft) to the ith receiver is defined by expression

$$D_i = |\mathbf{P}_i - \mathbf{M}| = \sqrt{(x_i - x)^2 + (y_i - y)^2 + (z_i - z)^2}. \tag{6.1}$$

To simplify the mathematical description of the task, we will place the beginning of the rectangular coordinate system to the point of installation of the antenna of one of the receivers, e.g., P_0. The master station is the best point to be used as that receiver. Then the distance from the aircraft to the receiver is determined by expression

$$D_0 = \sqrt{x^2 + y^2 + z^2}, \tag{6.2}$$

where x, y, z are the aircraft coordinates in the chosen coordinate system.

Distance D_i in expression (6.1) is the product of radio propagation velocity c and time T_i of signal propagation from the aircraft transponder to the ith MLAT ground-based receiver, i.e., $D_i = cT_i$. MLAT measures the differences $\Delta T_{i0} = T_i - T_0$ of times of arriving the signals at the pairs of ground stations (TDOA measurements).

Then the difference of distances from the aircraft to the pair of receivers located in points P_0 and P_i can be written as

$$\Delta D_{i0} = c\Delta T_{i0} = cT_i - cT_0 = D_i - D_0. \tag{6.3}$$

The TDOA measurements can be performed by calculating the cross-correlation function (CCF) of the signals received by the ground station pairs. The shaping of cross-correlation function for pulse signals is shown in Fig. 6.5. The interval between the times of signal arrival (300 ns) is measured by the shift of the maximum of correlation function on the reference time scale.

The locos of the points meeting condition (6.3) make up a hyperboloid in space. Using four receiving MLAT stations ($0 \leq i \leq 3$), it is possible to obtain the system of three nonlinear equations of (6.1) type with three unknowns (x, y, z). Having solved the equation system, we will obtain the unknown coordinates of the aircraft (transponder) in real time.

The MLAT system can use transponders whose signals contain information of height z. Then, to calculate the other two unknowns (x, y), it is necessary to have three or more receivers in points with known coordinates. It should be noted that the

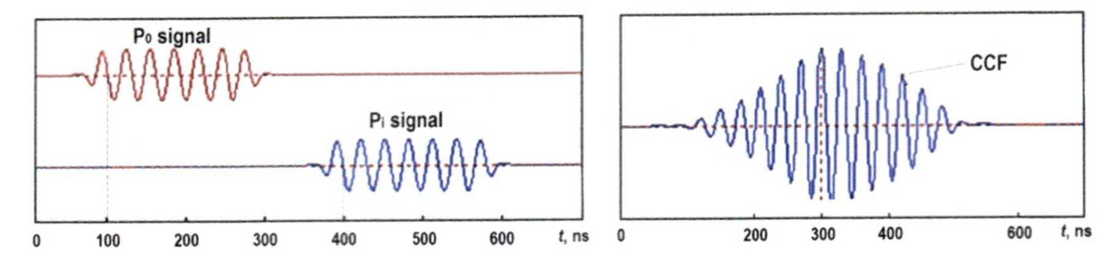

Fig. 6.5 Shaping of cross-correlation function for pulse signals

airborne transponders transmit information of barometric and not geometric altitude which should be taken into account when solving the task of aircraft positioning.

A greater number of receivers used for improvement of positioning accuracy causes a significant increase in computational costs since the system including more nonlinear equations of (6.1)–(6.2) type must be solved. Higher processor efficiency is therefore required.

The task can be simplified by using five or more receivers since it is possible to obtain the system of linear equations whose solution requires fewer computational costs.

6.3 Typical Variants of the MLAT Architecture

The MLAT architecture is specified by the number and mutual arrangement of ground stations as well as by the methods of measuring the TDOA of reply signals. In their turn, the ways of creating a time scale on which the TDOA measurements are performed (a common or distributed one) define the techniques of synchronizing the time scales of the receivers.

The number of the MLAT receiving stations is first of all specified by the required coverage area of the system. The larger the coverage area, the more stations should be included into the network. The number of ground stations used in the MLAT network is also influenced by the ground profile as additional stations may be needed to create a continuous and dependable field of vision in the given area. The mutual arrangement of ground stations influences the configuration of the coverage edges and accuracy of determining the aircraft coordinates.

The typical variants of MLAT stations arrangement are shown in Fig. 6.6 and the master station (with the MLAT controller) which is crucial for solving the task of aircraft positioning is, as a rule, in the center of the figure with the other system stations as its apexes. Such location of the master station will minimize the length of signal transmission lines and let know more precisely an additional signal delay when propagating via radio or other communication links from the receiving stations to the MLAT controller.

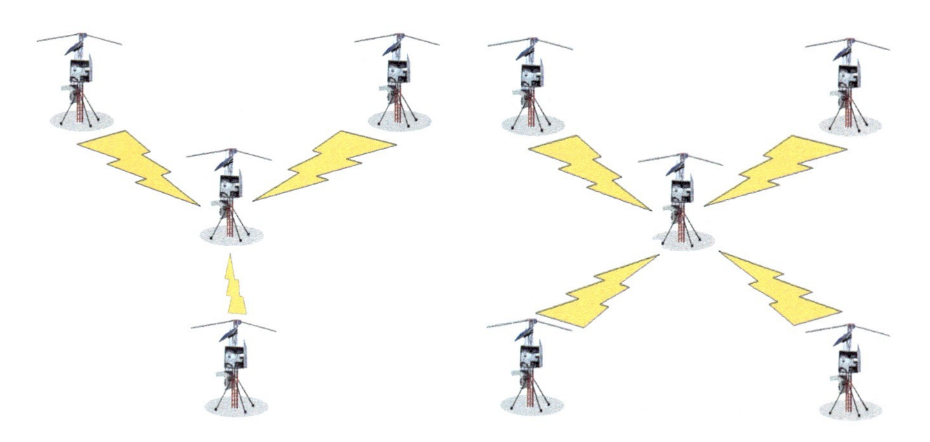

Fig. 6.6 Typical variants of arrangement of MLAT ground-based stations

The TDOA can be determined by calculating the cross-correlation function of signals received or by measuring the time of delay (TOA) of the signals received by ground stations and determining TDOA according to them.

The former technique can be used for signals of any type but its efficiency depends on the autocorrelation properties of the signal. The latter technique is suitable only for pulse signals which allow the pulse leading edge to be fixed with a high degree of accuracy.

In MLAT, the information about aircraft coordinates is contained in the difference of times of signal arrival at the ground stations of the system. To extract it, the times should be reduced to a unified time scale, i.e., the measurements should be synchronized. It is essential to take into account additional delays of signals while they are processed in the receiving paths and when the information is transmitted from the receiving stations to the master one.

The MLAT measurements can be reduced to a unified scale by the following means:

1. The reference time scale is created at one station (in this case at the master station) which obtains the signals received by the other stations of the system and where they are processed by the MLAT controller. On this scale, the TOA are fixed with the account of the delay caused by the distance from each network station to the master one (by size of the system bases).

Instability of the reference time scale created at the master station by means of a crystal oscillator is at most 1 ns.

If the unified time scale is used, the MLAT receiving equipment is rather simple and its function is to receive and retranslate the received reply signals to the master station where the measurements are bound to the time and TDOA are formed and processed. However, the aircraft positioning accuracy is greatly dependent on the delay of signals when they are transmitted via the communication channels and processed in the master station processor.

2. The distributed time scales are created at each station of the network and their mutual synchronization is provided, for example, by GNSS time or by other means. But again, the information is processed at the master station with account of the sizes of the system bases.

Need for synchronization of the distributed time scales causes the receiver circuits to get sophisticated as each of them processes the received reply signals for obtaining the information about the delay (TOA).

To synchronize the time scales, one can use such methods as transponder synchronization and external GNSS synchronization.

The transponder synchronization requires reference time signals transmitted by the reference transponder of the MLAT ground stations. This kind of synchronization can therefore be called internal one. The reference transponder does not have to be located at the master station; it must be situated within line of sight of all system stations to enable them to receive consistently synchronization signals.

The method should be used for small-sized station bases to avoid the influence of wave propagation conditions on the additional delay of the synchronization signal in its propagating from the reference transponder to the synchronized receivers.

The method has an important advantage. As the synchronization signal and the aircraft reply signals pass through the same processing path in the receiver, the additional group delay caused by the path is eliminated from the TOA measurements on the receiver time scale.

In expanded MLAT systems, several synchronizing transponders can be used given that each pair of receivers that uses common reference signals with each other pair.

External GNSS synchronization of distributed time scales is an external source of common reference time for each receiver as time in GNSS is sufficiently accurate (the difference with UTC is no more than 100 ns). It is possible to use a GNSS-synchronized reference crystal oscillator and to synchronize the MLAT receivers with accuracy up to 10–20 ns.

As in MLAT, the time difference of signal arrival at the stations is measured; the error of synchronization with UTC time equal for all receivers is compensated in subtracting TOA.

The method of external GNSS synchronization is much easier than creation of a unified time scale or reference transponder synchronization. It allows any digital channels to be used for transmitting a synchronization signal and does not require placement of the reference transponder and synchronized receivers within the line of sight. However, to provide reliable synchronization which depends on GNSS integrity, GNSS receivers with RAIM function should be used.

When autonomous GNSS synchronization of receivers is insufficiently accurate, one can use the method of synchronization on common visible GNSS satellites which are within the visibility zone of the network receivers. Since this method is similar to the differential mode of GNSS measurements, it eliminates the influence of causes of common errors, e.g., ephemeris errors and ionospheric disturbances, and provides a much more accurate synchronization. The synchronization accuracy is several fractions of a nanosecond.

Corrections for time-scale synchronization are made either directly in TOA measurements for each receiver or when they enter the MLAT controller at the master station. In both cases, there is no need to have a GNSS receiver at the master station as the TOA measurements are correlated with the time scale at the receiving stations.

Due to processing the signals from all common visible satellites, the RAIM function is performed which controls the integrity of synchronization data quality among receiving stations and provides integrity estimate with higher degree of confidence.

The MLAT systems can be passive or active when their ground part includes one or several transmitting devices whose signals are received by the airborne transponder and cause formation and radiation of a reply signal.

For passive MLAT systems to operate, one of the following requirements should be met:

the aircraft is equipped with a self-generating Mode S transponder;
the aircraft is equipped with a Mode A/C transponder and is within the coverage of one or several SSR interrogators;
the aircraft is equipped with a Mode A/C transponder and is within the coverage of aircraft equipped with ACAS interrogators.

This means that the passive MLAT systems can be applied in airspace areas with high density of air traffic equipped with ACAS and/or secondary radio location means as well as in the areas where Mode S is obligatory to be used.

The passive MLAT systems are difficult to be used if aircraft is equipped only with Mode A/C transponders and fly at low altitudes as in this case, the airborne reception of interrogation signals from ground-based SSR is less possible.

Active MLAT systems can solve the same tasks as the passive systems do but they are capable to initiate themselves reply signals of airborne transponders. A MLAT interrogator is much simpler than a SSR interrogator as it can have lower power and a non-directional antenna.

Thus, the passive MLAT systems are preferably used along the air routes where there is a ground-based SSR surveillance field. The active MLAT systems are preferably used in the terminal area as the use of their own interrogator provides reception of reply signals from aircraft descending beyond the ground-based SSR coverage. Besides, due to increasing the frequency of reply signal formation (when interrogating selectively individual aircraft), it is possible to expedite updating of information about the aircraft coordinates and accuracy of their determination.

6.4 Factors Influencing the MLAT Accuracy

Positioning accuracy in MLAT depends on a number of factors, the most important being TDOA measurement accuracy, location of ground receiving stations, and aircraft position relative to these stations.

The accuracy of TDOA measurements depends on the used method of measurements, accuracy of time-scale synchronization, parameters of the reply signal, performances of receiving devices (internal noise, instability of group delay of a signal in processing paths), state of radio channels of reply signal propagation, and data transmission among the MLAT stations (instability of the speed of radio-wave propagation).

The mentioned constituents of the resultant error of TDOA measurements have a random character. To decrease the resultant error, in addition to the measures on decrease of each constituent, the structure redundancy of the system is used (the number of receivers is considerably larger than minimal three). It also can increase the system coverage. However, in this case, the MLAT accuracy characteristics will depend on the aircraft flight altitude since the number of ground stations capable of receiving reply signals is determined by conditions of radio line of sight of the aircraft.

To estimate the influence of aircraft position relative to the MLAT receiving stations on its positioning accuracy, the concept of ***Geometric Dilution of Precision (GDOP)*** or ***DOP*** is used (like in satellite radio navigation systems).

The GDOP expression can be produced by Taylor series expansion (6.3) in the neighborhood of the true values of aircraft coordinates x^0, y^0, z^0. Limiting ourselves to the linear terms of the series, we will obtain

$$\begin{aligned}\delta D_{i0} = {} & \left.\frac{\partial D_i}{\partial x}\right|_{\substack{x=x^0\\ y=y^0\\ z=z^0}} \delta x + \left.\frac{\partial D_i}{\partial y}\right|_{\substack{x=x^0\\ y=y^0\\ z=z^0}} \delta y + \left.\frac{\partial D_i}{\partial z}\right|_{\substack{x=x^0\\ y=y^0\\ z=z^0}} \delta z \\ & + \left.\frac{\partial D_0}{\partial x}\right|_{\substack{x=x^0\\ y=y^0\\ z=z^0}} \delta x + \left.\frac{\partial D_0}{\partial y}\right|_{\substack{x=x^0\\ y=y^0\\ z=z^0}} \delta y + \left.\frac{\partial D_0}{\partial z}\right|_{\substack{x=x^0\\ y=y^0\\ z=z^0}} \delta z, \end{aligned} \tag{6.4}$$

where δD_{i0} is a measurement error of difference of range from the aircraft to the ith receiver and master stations; x^0, y^0, z^0 are the true aircraft coordinates; δx, δy, δz are errors of aircraft coordinates determination.

Having taken partial derivatives, after grouping the similar terms, we will write expression (6.4) as

$$\begin{aligned}\delta D_{i0} = {} & \left[\frac{x^0 - x_i}{D_i} - \frac{x^0 - x_0}{D_0}\right]\delta x + \left[\frac{y^0 - y_i}{D_i} - \frac{y^0 - y_0}{D_0}\right]\delta y \\ & + \left[\frac{z^0 - z_i}{D_i} - \frac{z^0 - z_0}{D_0}\right]\delta z. \end{aligned} \tag{6.5}$$

Since in the MLAT system it is necessary to measure the range differences at least for three pairs of stations to determine the aircraft coordinates, we will introduce vector $\mathbf{\delta Z}^{\mathrm{T}} = |\delta D_{10}, \delta D_{20}, \delta D_{30}|$ containing difference measurement errors of range to the master and each of three receiving stations.

We will also introduce vector $\mathbf{\delta X}^{\mathrm{T}} = |\delta x, \delta y, \delta z|$ containing the errors of aircraft coordinates' measurements.

Then the system of three equations of type (6.5) for errors of TDOA measurements in the MLAT system can be written in the vector-matrix form

$$\mathbf{\delta Z} = \boldsymbol{H}\mathbf{\delta}\boldsymbol{X}, \tag{6.6}$$

where $\boldsymbol{H}$ is a matrix of directional cosines of "ground-based receiver—aircraft" range lines in the coordinate system $OXYZ$ which is written as

$$\boldsymbol{H} = \left\| \begin{bmatrix} \frac{x^0 - x_1}{D_{11}} - \frac{x^0 - x_0}{D_0} \\ \frac{x^0 - x_2}{D_2} - \frac{x^0 - x_0}{D_0} \\ \frac{x^0 - x_3}{D_3} - \frac{x^0 - x_0}{D_0} \end{bmatrix} \begin{bmatrix} \frac{y^0 - y_1}{D_1} - \frac{y^0 - y_0}{D_0} \\ \frac{y^0 - y_2}{D_2} - \frac{y^0 - y_0}{D_0} \\ \frac{y^0 - y_3}{D_3} - \frac{y^0 - y_0}{D_0} \end{bmatrix} \begin{bmatrix} \frac{z^0 - z_1}{D_1} - \frac{z^0 - z_0}{D_0} \\ \frac{z^0 - z_2}{D_2} - \frac{z^0 - z_0}{D_0} \\ \frac{z^0 - z_3}{D_3} - \frac{z^0 - z_0}{D_0} \end{bmatrix} \right\|.$$

From expression (6.6), it is possible to find the variance matrix of errors of range difference measurements $\boldsymbol{D}_{\delta Z} = D\{\mathbf{\delta Z}\} = M\{\mathbf{\delta Z}\mathbf{\delta Z}^{\mathrm{T}}\}$, where "T" is a transposing symbol, $M\{o\}$ is a mathematical expectation symbol. Subject to the transposing rules, we will obtain

$$\boldsymbol{D}_{\delta Z} = M\{\boldsymbol{H}\,\mathbf{\delta X}\mathbf{\delta X}^{\mathrm{T}}\boldsymbol{H}^{\mathrm{T}}\} = \boldsymbol{H}M\{\mathbf{\delta X}\mathbf{\delta X}^{\mathrm{T}}\}\boldsymbol{H}^{\mathrm{T}} = \boldsymbol{H}\boldsymbol{D}_{\delta X}\boldsymbol{H}^{\mathrm{T}}.$$

From this, we will find the variance matrix of errors of aircraft coordinates determination

$$\boldsymbol{D}_{\delta X} = \boldsymbol{H}^{-1}\boldsymbol{D}_{\delta Z}(\boldsymbol{H}^{\mathrm{T}})^{-1},$$

where (-1) is a matrix inversion symbol.

Considering the range difference measurements as uniformly precise ($\sigma_{\delta D1} = \sigma_{\delta D2} = \sigma_{\delta D3} = \sigma_{\delta D}$), we will obtain

$$\boldsymbol{D}_{\delta X} = \boldsymbol{H}^{-1}\left(\boldsymbol{H}^{\mathrm{T}}\right)^{-1}\sigma_{\delta D}^{2} = \left(\boldsymbol{H}^{\mathrm{T}}\boldsymbol{H}\right)^{-1}\sigma_{\delta D}^{2}.$$

Aircraft positioning accuracy can be defined as a sum of diagonal elements of matrix $\boldsymbol{D}_{\delta X}$. Then

$$\sigma_{\delta x}^{2} + \sigma_{\delta y}^{2} + \sigma_{\delta z}^{2} = \mathrm{tr}\left[\left(\boldsymbol{H}^{\mathrm{T}}\boldsymbol{H}\right)^{-1}\right]\sigma_{\delta D}^{2} = \mathrm{GDOP}\cdot\sigma_{\delta D}^{2},$$

where

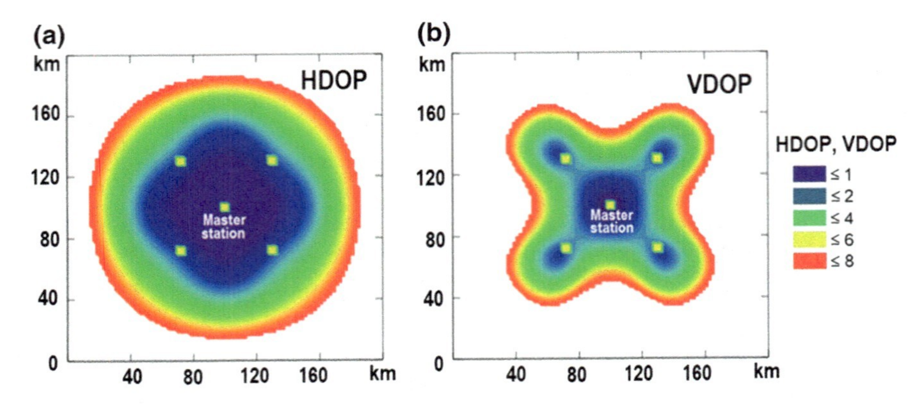

Fig. 6.7 Accuracy fields of MLAT consisting of five stations: **a** horizontal coordinates determination (HDOP); **b** height determination (VDOP)

$$\mathrm{GDOP} = \sqrt{\mathrm{tr}\left[\left(\boldsymbol{H}^{\mathrm{T}}\boldsymbol{H}\right)^{-1}\right]}$$

is a coefficient correlating the aircraft positioning error dispersion and the error dispersion of measuring the difference of ranges to the aircraft.

The following GDOP constituents for the MLAT system are distinguished:

horizontal (HDOP)—characterizes the accuracy of aircraft positioning in horizontal plane;
vertical (VDOP)—characterizes the accuracy of aircraft positioning in vertical plane;
spatial or position (PDOP)—characterizes the accuracy of aircraft positioning in space.

Distribution of the accuracy field for determination of the aircraft horizontal coordinates (HDOP values) and the accuracy field for its height determination (VDOP values) is considerably different. As an example, you can see in Fig. 6.7 the accuracy fields of the MLAT system with ground stations making up a square with the master station in the center.

The HDOP and VDOP values depend on the aircraft position relative to the ground MLAT stations and the system geometry—sizes of bases (Fig. 6.8).

The MLAT peculiarity is that the accuracy field for horizontal coordinates determination is fairly uniform, while the configuration of the accuracy field for the height determination greatly depends on the aircraft flying altitude (Fig. 6.9), number of ground stations (Fig. 6.10) and location of master station (Fig. 6.11).

In the whole, the best positioning accuracy is when the aircraft is inside the figure made up by the MLAT ground stations and their bases. Removal from baselines outside considerably reduces the positioning accuracy. However, the distribution of the MLAT accuracy field is much better than aircraft positioning by radar data when the accuracy gets much worse with removing from the radar antenna (rho-theta method of coordinates determination).

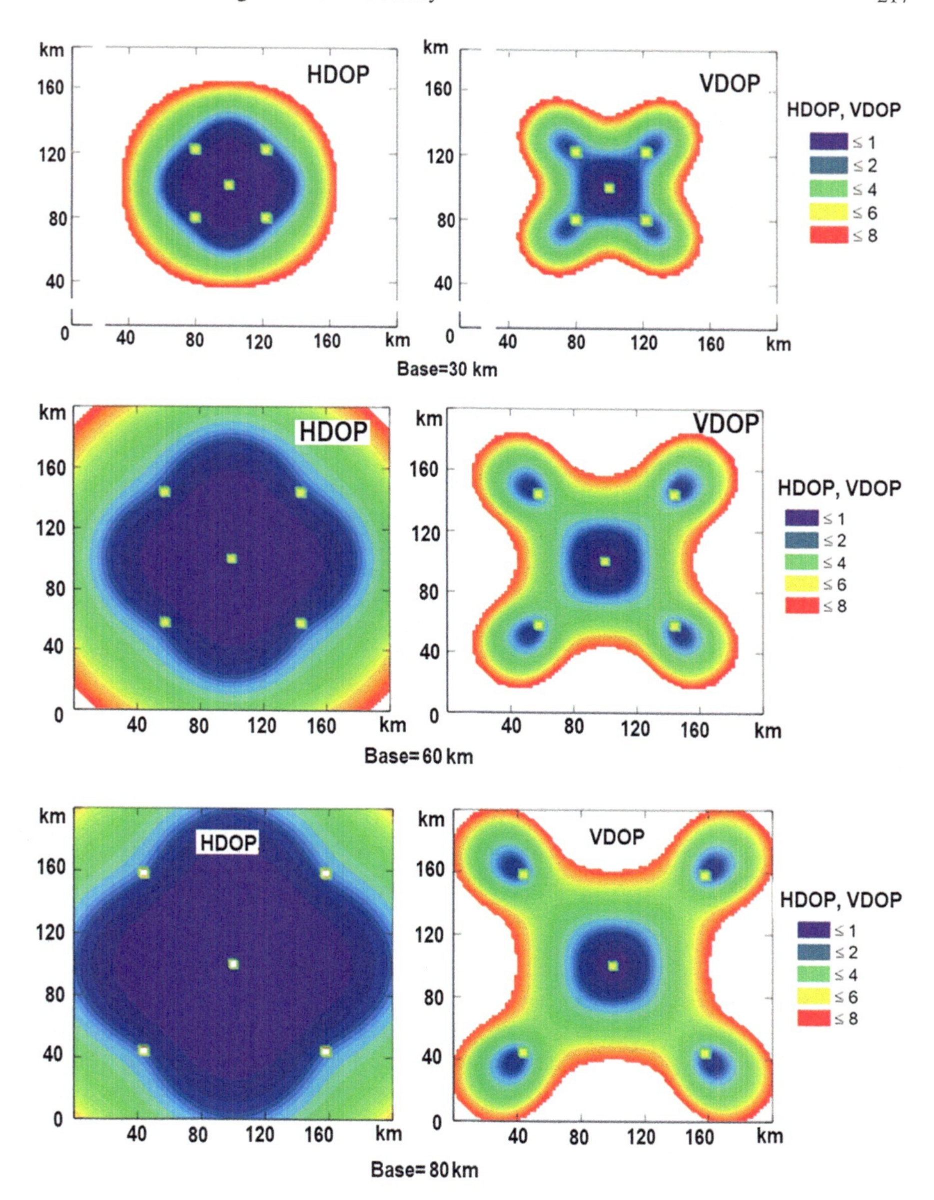

Fig. 6.8 Accuracy fields of horizontal coordinates and height measurements depending on the size of MLAT station bases

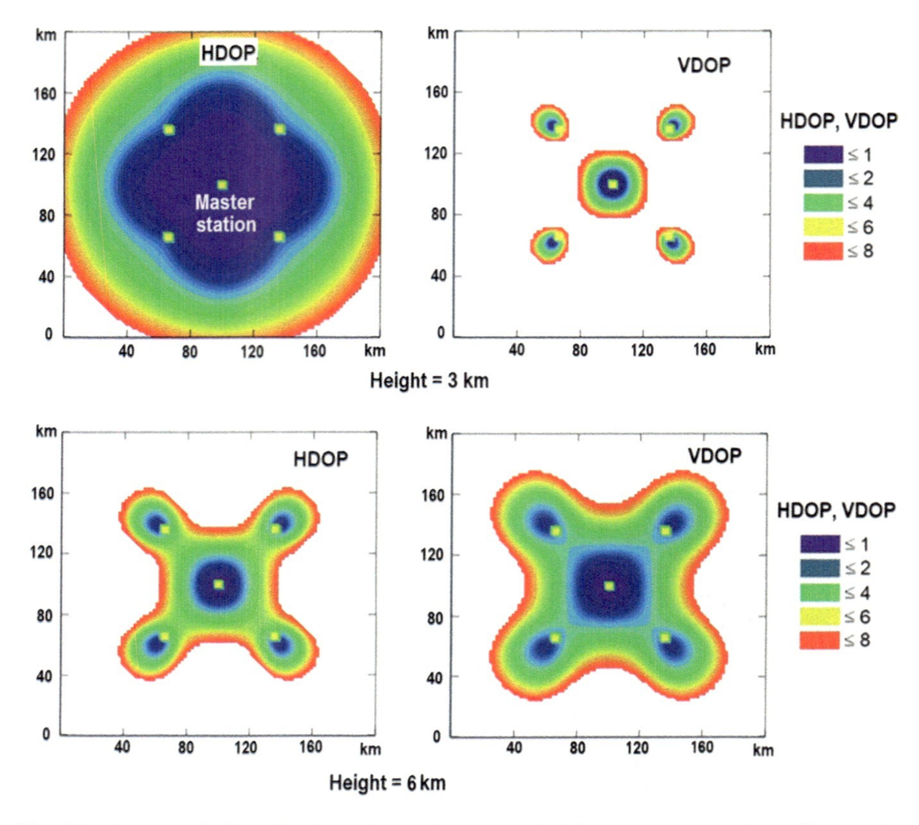

Fig. 6.9 Accuracy fields of horizontal coordinates and height measurements depending on the aircraft flying altitude

The configuration and size of the MLAT accuracy field are dependent on the system operation mode (active or passive).

In Sect 1.7.5, the author considers a technique of construction and analyzes the working area of classical differential distance long-range navigation systems Chayka (Russia) and LORAN-C (USA). As the systems have large base sizes (up to 500–700 km), the navigation task is usually solved using two pairs of stations. In the vicinity of station bases or along the outbase lines, therefore, there is no working area of the system or the system accuracy is unacceptably low. The reason is that in these sections of the working area, the GDOP of the differential distance long-range navigation system determined by expression

$$K(\psi_{B1},\psi_{B2},\alpha_M)=\frac{\sqrt{\sin^2\frac{\psi_{B2}}{2}+\sin^2\frac{\psi_{B1}}{2}}}{2\sin\alpha_M\sin\frac{\psi_{B1}}{2}\sin\frac{\psi_{B2}}{2}}$$

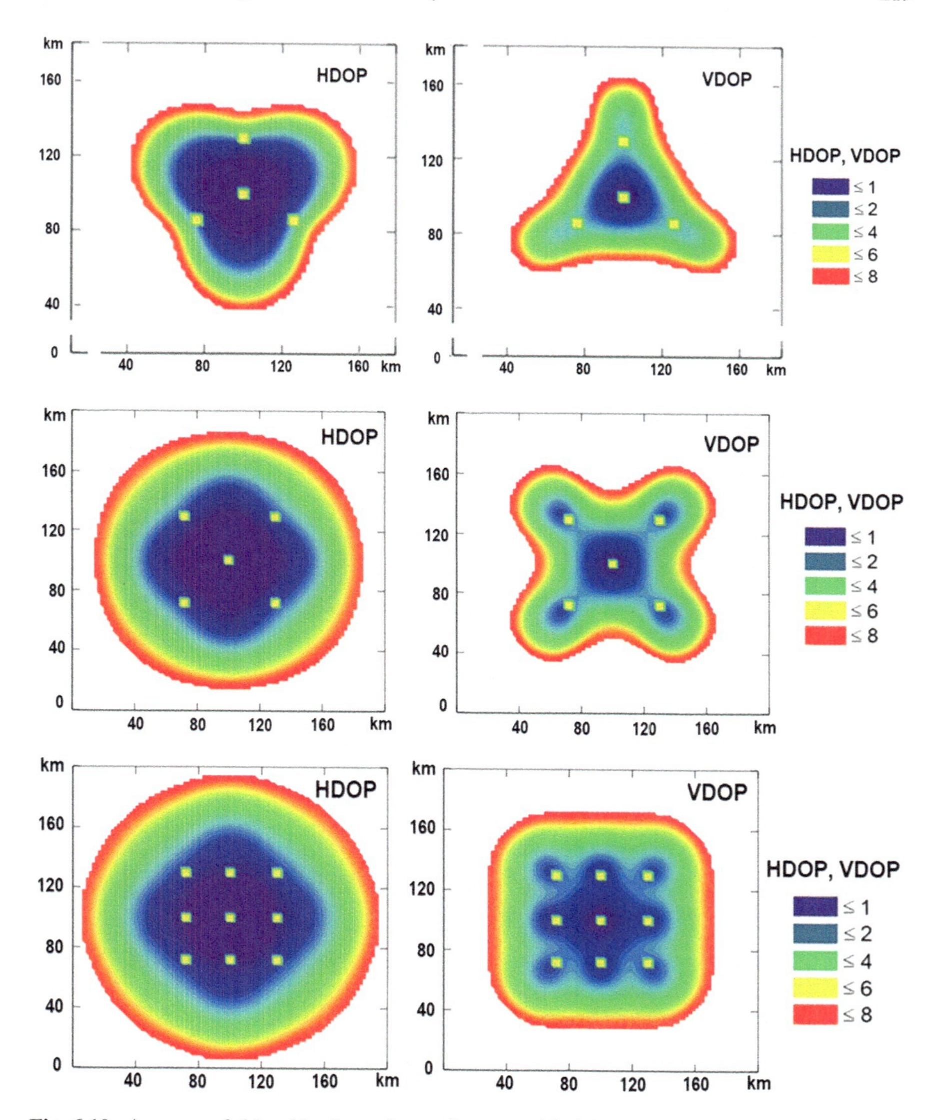

Fig. 6.10 Accuracy fields of horizontal coordinates and height measurements depending on the number of MLAT stations

is considerably degraded because datum angles ψ_{B1}, ψ_{B2} are close to zero or the lines of position intersect at angle α_M close to zero (ref. Fig. 1.35).

This problem is not relevant for MLAT due to relatively small base sizes and, hence, ability to use excessive TDOA measurements almost of all pairs of ground stations for solving the navigation task.

However, in the MLAT system, like in a classical differential distance long-range navigation system, the area of working zone will decrease with degrading the accu-

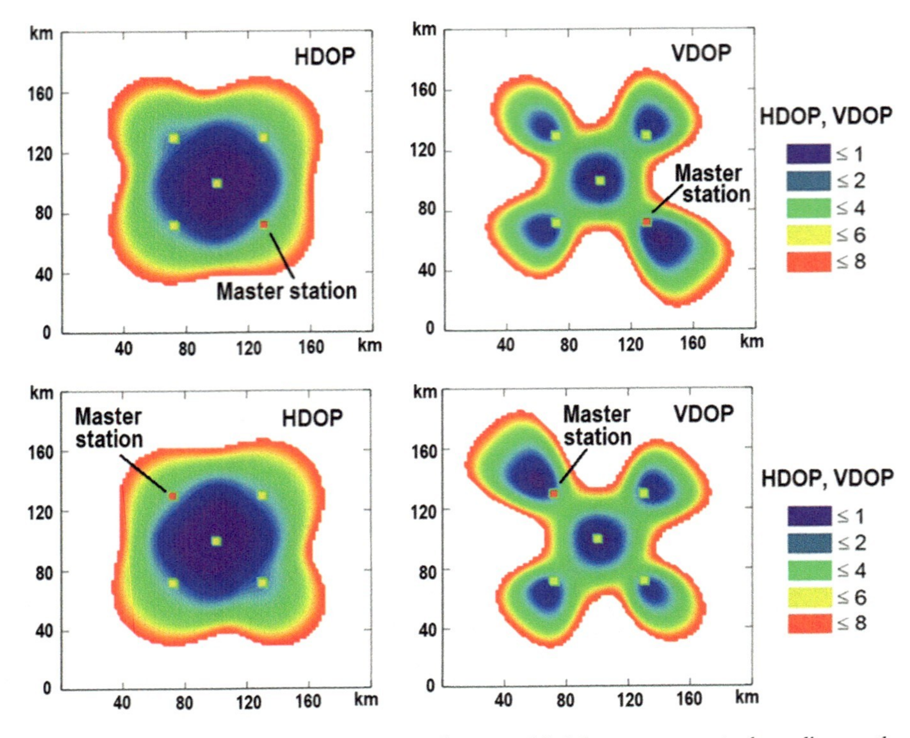

Fig. 6.11 Accuracy fields of horizontal coordinates and height measurements depending on the master station position

racy of TDOA measurements and with increasing the demands to the accuracy of aircraft coordinates determination.

6.5 Requirements to the MLAT Performances

As the MLAT system is designed to solve surveillance tasks, it has to meet the requirements of the similar systems according to the ICAO regulatory documents. Accordingly, the airdrome MLAT system has to meet the requirements of Annex 10 to the Convention on International Civil Aviation. Vol. IV. Surveillance and Collision Avoidance Systems.

On the basis of that and a number of other ICAO documents, minimal requirements are developed for the wide area MLAT systems designed for monitoring the airspace along the airways and in the airdrome area. For the MLAT systems used in the Russian Federation, "Certification requirements (basis) to wide area and airdrome multiposition surveillance systems" were approved in 2016 and their provisions are consistent with the ICAO requirements.

In accordance with the certification requirements, the MLAT system has to receive and process information from an aircraft equipped with the transponders operating in A/C and S Modes and with the equipment of generation of extended squitters (ADS-B 1090 ES) as well as to transmit interrogations to the airborne transponders. The MLAT equipment should include at least four receiving stations and an active MLAT system should have at least one interrogator.

Frequency of 1,030 ± 0.01 MHz is assigned for operating the interrogation channel of an active MLAT, while the reply signals are received at the frequencies of 1,090 ± 3 MHz (in RBS mode) and 1,090 ± 1 MHz (in S and ADS-B 1090 ES modes). The vertically polarized signals are used in the interrogation and reply channels.

The power of radiated interrogation signals in the active MLAT system should be adjustable and its capacity should be a configurable parameter. The MLAT system should provide detection, identification, and guidance of at least 250 targets in the assigned coverage.

In accordance with the requirements laid down for the means of radio-technical support of flights, the MLAT system should have a remote control and monitoring system which provides:

continuous monitoring of technical state and control of operational modes of the system and its components from the operator's work station;
automatic configuration of the system in case if any of its redundant elements fails;
automatic indication of the MLAT current configuration, changes of the technical state, and operational modes of the equipment;
reception and indication of two messages of functional check—"Norm" and "Fault";
two operational modes: «Operation» и «Maintenance».

Technical specifications for the wide area and airdrome MLAT systems are distinguished by the parameters which define the purpose and application peculiarities of the systems.

So, for a wide area MLAT system, the time interval for updating the information about the aircraft position should not exceed 8 s for the route and 4 s for the approach zone. For an airdrome MLAT system, the time interval for updating the information should not exceed 1 s with a probability of 0.7 in the apron zone, 0.5 in the stand zone, 0.95 in the maneuvering zone or at a distance up to 9.6 km from the runway threshold. Notice that the required probability of updating the information increases with the growth of the aircraft dynamic capabilities.

For the wide area MLAT system, the standard error of determining the horizontal position of an aircraft should not exceed 350 m along the route and 150 m in the terminal area. For the airdrome MLAT system, much harder requirements on accuracy are put forward: 7.5 m ($P = 95\%$) and 12 m ($P = 99\%$) for the airdrome maneuvering area, 20 m (mean accuracy for 5 s) for the stand zone, 20 m ($P = 95\%$) for an area of less than 4.6 km from the runway threshold for air targets, and 40 m ($P = 95\%$) for an area of 4.6–9.6 km in radius from the runway threshold for air targets. It should be noted that the requirements to the aircraft positioning accuracy are increased as the aircraft approaches the runway threshold.

There is a requirement for the MLAT system to provide the output information about the coordinates of aircraft and other airdrome vehicles in the airdrome Cartesian (coordinates x, y, z) and geodetic (latitude and longitude) coordinate systems. MLAT should also determine the pressure altitude or the actual height of an aircraft in the airdrome Cartesian and geodetic coordinate systems. The geodetic coordinates should be provided in PZ-90.2 (for MLAT used in Russia) and WGS-84 coordinate systems.

Besides, MLAT should provide information about the positioning accuracy in the specified coordinate systems (RAIM function).

Nowadays, the MLAT systems are in use worldwide. In Russia, such systems have recently been intensively developed and implemented (wide area MLAT "MERA," airdrome MLAT of "TETRA" family, integrated MLAT "Al'manah"), first of all at the large airports with high density of air and airdrome traffic.

So, since 2013, airdrome MLAT ERA MSS have been operated at Sochi and Domodedovo airports; in 2018, "Al'manah" system was introduced into service at Pulkovo airport, and "Tetra" system was introduced into service at Sheremetyevo and Vnukovo airports.

It is expected that the MLAT systems will be installed at all airports which have runways certified under Category II and are able to serve flights day and night in intricate meteorological conditions.

"Al'manah" (developed by "CRTS" Research and development enterprise) is Russia's first and the world's second integrated multiposition surveillance system performing the functions of both airdrome and wide area means of surveillance. The system operates in all surveillance modes including ADS-B 1090 ES which is recommended by ICAO as a common standard for the whole world's airspace. Besides, "Al'manah" tracks the coordinates of aircraft and other vehicles on the airfield and on the glide path.

The MLAT system includes the following objects:

receiving stations (Fig. 6.12a) designed for reception and decoding of airborne transponder signals as well as for transmission of the data about the times of arriving the signals at the master station. ADS-B stations can be used as receiving ones;
transmitting stations (interrogators) which allow the MLAT system to operate in the active mode;
test-and-reference transponders used for testing the system and for internal synchronization of the distributed time scales of receivers;
a MLAT server and a MLAT controller (Fig. 6.12b) which receive information from ground stations, calculate TDOA and aircraft coordinates, control the system, and transmit the output information for users in an appropriate format.

Figure 6.13a, b shows the design of antennas for receiving stations of the MLAT. Besides, to improve the safety and monitoring of airdrome traffic, small-sized radio beacon antennas (Fig. 6.13c) are used. They are installed on ground vehicles and transmit signals (ADS-B squitters) of their precise location.

Additionally, the system can include stations monitoring the reception and interference.

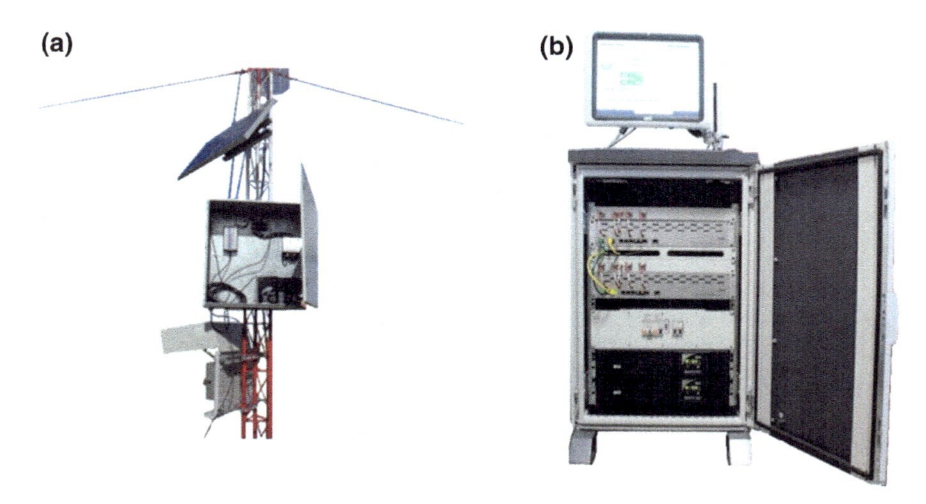

Fig. 6.12 Structural design of the MLAT objects. *Source* https://itk-mdl.asutk.ru/upload/iblock/4c1/МПСН%20Альманах.pdf

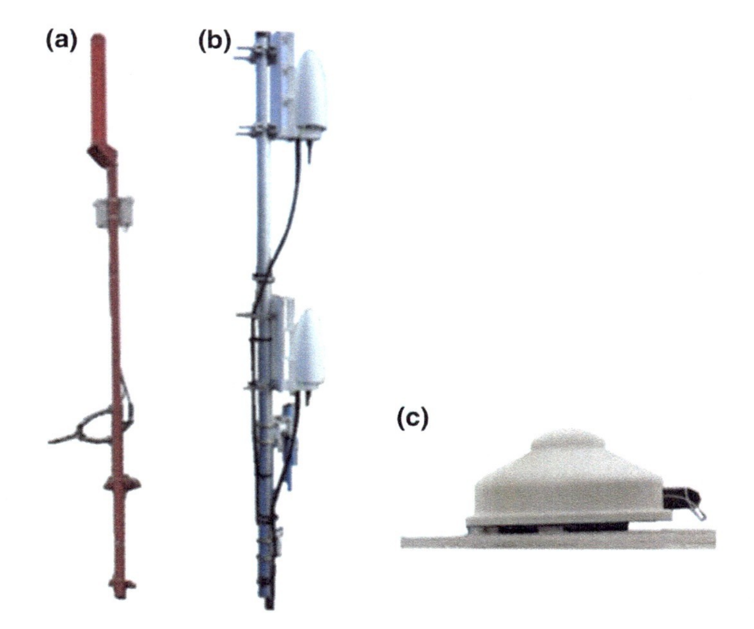

Fig. 6.13 Antennas of the MLAT objects

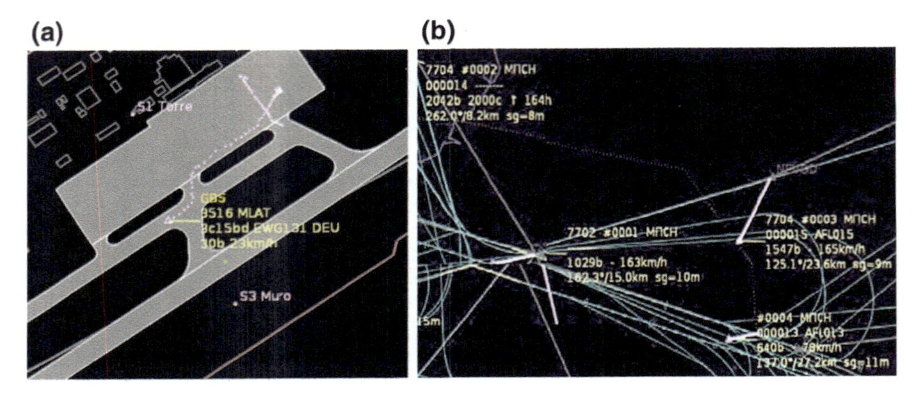

Fig. 6.14 MLAT tracks. *Source* https://itk-mdl.asutk.ru/upload/iblock/4c1/МПСН%20Альманах.pdf

The system can use different ways of synchronizing the spaced time scales of the receivers: according to the test-and-reference interrogator, on the GLONASS and GPS time scales and signals.

Technical solutions implemented in the receivers allow the system to operate:

in the active and passive modes;
in the mode of a wide area MLAT system creating a surveillance field in the airdrome zone with the radius of 100 km;
in the mode of an airdrome MLAT system including monitoring of movements of aircraft and vehicles on the airdrome surface;
in the mode of controlling the aircraft altitude hold necessary for providing the RVSM technology (reduces intervals of vertical separation).

The system capacity is 250 units but technically the system can provide airdrome monitoring with up to 1,000 objects in the surveillance field.

To organize indication on ATC work stations, the "Al'manah" system is interconnected with the complex automatization systems "Vega" and "Galaktika." In order to provide the aircraft safety on approach, taxiing, taking off, and especially in poor visibility or poor weather conditions, the system allows a controller to see movement and position tracks of aircraft (Fig. 6.14b) and ground vehicles (Fig. 6.14a) on the movement monitoring screen. The screen also displays a tracking form containing the callsign and speed of the aircraft or vehicle.

The MLAT advantages are:

the use of the signals already existing in the secondary radar system for solving a new task—aircraft coordinates determination, thus not requiring installation of any additional equipment on board the aircraft;
sufficient high rate of information update (1 s);
sufficient high accuracy of aircraft positioning;

possibility to create a surveillance field in the specified zone of airspace in challenging conditions—at low altitudes, in mountains, etc.;
simplicity of maintenance and operation, low power consumed by the system, no need in constructing separate structures for ground stations;
possibility to perform the functions of both airdrome and en route surveillance means simultaneously.

The MLAT disadvantages are:

dependence of positioning accuracy on the location of ground stations and aircraft position relative to them;
necessity to synchronize the time scales of receivers;
necessity to have infrastructure for placement of MLAT objects as well as the data transmission channels between the receiving stations and the MLAT controller.

In the future, the surveillance task will be solved by means of ADS-B system. Therefore, the MLAT system can currently be considered as an augmentation of a radar surveillance system (PSR and SSR) for transition to ADS-B surveillance. The MLAT system can further be used as a redundant system and as a means of ADS-B integrity control.

6.6 Summary

MLAT systems proved its high efficiency as a mean which completes, and in some cases, replaces traditional means of surveillance (radio locators). That is why such systems find wide usage although not exclusively for solving the surveillance tasks in the airfields' area but also on air routes.

Unique feature of MLAT systems is the usage of navigation methods for receiving the information about aircrafts' coordinates from signals of airborne transponders—elements of secondary radio location systems.

This chapter provides information that helps to understand main characteristics, operation principles, versions of the ground stations' network construction, and also to estimate the influence of MLAT system's architecture on its operating area configuration. At analyzing the MLAT operating area, author used the results of researches which were performed under his supervision.

6.7 Further Reading

Information on ICAO requirements to the parameters and characteristics of MLAT can be found in [1].

In [2, 3], common characteristics of the system and aspects of its use are given.

A lot of useful information which include information of exploratory type can be found in [4].

There is a wealth of information on MLAT at https://en.wikipedia.org/wiki/ multilateration.

Information on main characteristics and constructional special aspects of Russian-made MLAT systems can be found on the Web sites of manufacturers. For example http://npp-crts.ru/en/, http://www.vniira.ru/doc/catalogue/1179_en.pdf.

References

1. Aeronautical Surveillance Manual, Doc. 9924, 2nd edn. ICAO, Montreal (2017). Download from https://standards.globalspec.com/std/10174090/ICAO%2099242
2. Multilateration (MLAT) Concept of Use, 1.0 edn. ICAO Asia and Pacific Office, September 2007. Download from https://www.icao.int/APAC/Documents/edocs/mlat_concept.pdf
3. Multilateration. Executive reference guide. CREATIVERGE, ERA corporation. Download from http://www.multilateration.com/downloads/MLAT-ADS-B-Reference-Guide.pdf
4. Neven WHL, Quilter TJ, Weedon R, Hogendoorn RA (2005) NLR-CR-2004-472 wide area multilateration report on EATMP TRS 131/04 version 1.1. Eurocontrol. Download from https://www.eurocontrol.int/sites/default/files/publication/files/surveilllance-report-wide-area-multilateration-200508.pdf

Printed by Printforce, the Netherlands